# SI units

| Physical quantity | Old unit | Value in SI units |
|---|---|---|
| energy | calorie (thermochemical) | 4·184 J (joule) |
| | *electronvolt—eV | $1·602 \times 10^{-19}$ J |
| | *electronvolt per molecule | 96·48 kJ mol$^{-1}$ |
| | erg | $10^{-7}$ J |
| | *wave number—cm$^{-1}$ | $1·986 \times 10^{-23}$ J |
| entropy (S) | eu = cal g$^{-1}$ °C$^{-1}$ | 4184 J kg$^{-1}$ K$^{-1}$ |
| force | dyne | $10^{-5}$ N (newton) |
| pressure (P) | atmosphere | $1·013 \times 10^{5}$ Pa (pascal), or N m$^{-2}$ |
| | torr = mmHg | 133·3 Pa |
| dipole moment (μ) | debye—D | $3·334 \times 10^{-30}$ C m |
| magnetic flux density (H) | *gauss—G | $10^{-4}$ T (tesla) |
| frequency (ν) | cycle per second | 1 Hz (hertz) |
| relative permittivity (ε) | dielectric constant | 1 |
| temperature (T) | *°C and °K | 1 K (kelvin); 0 °C = 273·2 K |

(* indicates permitted non-SI unit)

Multiples of the base units are illustrated by length

| fraction | $10^{9}$ | $10^{6}$ | $10^{3}$ | 1 | $(10^{-2})$ | $10^{-3}$ | $10^{-6}$ | $10^{-9}$ | $(10^{-10})$ | $10^{-12}$ |
|---|---|---|---|---|---|---|---|---|---|---|
| prefix | giga- | mega- | kilo- | metre | (centi-) | milli- | micro- | nano- | (*ångstrom) | pico- |
| unit | Gm | Mm | km | m | (cm) | mm | μm | nm | (*Å) | pm |

**The fundamental constants**

| | | |
|---|---|---|
| Avogadro constant | $L$ or $N_A$ | $6·022 \times 10^{23}$ mol$^{-1}$ |
| Bohr magneton | $\mu_B$ | $9·274 \times 10^{-24}$ J T$^{-1}$ |
| Bohr radius | $a_0$ | $5·292 \times 10^{-11}$ m |
| Boltzmann constant | $k$ | $1·381 \times 10^{-23}$ J K$^{-1}$ |
| charge of a proton (charge of an electron = $-e$) | $e$ | $1·602 \times 10^{-19}$ C |
| Faraday constant | $F$ | $9·649 \times 10^{4}$ C mol$^{-1}$ |
| gas constant | $R$ | $8·314$ J K$^{-1}$ mol$^{-1}$ |
| nuclear magneton | $\mu_N$ | $5·051 \times 10^{-27}$ J T$^{-1}$ |
| permeability of a vacuum | $\mu_0$ | $4\pi \times 10^{-7}$ H m$^{-1}$ or N A$^{-2}$ |
| permittivity of a vacuum | $\varepsilon_0$ | $8·854 \times 10^{-12}$ F m$^{-1}$ |
| Planck constant | $h$ | $6·626 \times 10^{-34}$ J s |
| (Planck constant)/$2\pi$ | $\hbar$ | $1·055 \times 10^{-34}$ J s |
| rest mass of electron | $m_e$ | $9·110 \times 10^{-31}$ kg |
| rest mass of proton | $m_p$ | $1·673 \times 10^{-27}$ kg |
| speed of light in a vacuum | $c$ | $2·998 \times 10^{8}$ m s$^{-1}$ |

$\ln 10 = 2·303$    $\ln x = 2·303 \lg x$    $\lg e = 0·4343$    $\pi = 3·142$
$R \ln 10 = 19·14$ J K$^{-1}$ mol$^{-1}$    $RTF^{-1} \ln 10 = 59·16$ mV at 298·2 K

# Oxford Chemistry Series

General Editors
**P. W. ATKINS     J. S. E. HOLKER     A. K. HOLLIDAY**

# Oxford Chemistry Series

GORDON HUGHES

DEPARTMENT OF INORGANIC, PHYSICAL AND INDUSTRIAL CHEMISTRY,
LIVERPOOL UNIVERSITY

# Radiation chemistry

Clarendon Press · Oxford · 1973

*Oxford University Press, Ely House, London W.1*

GLASGOW   NEW YORK   TORONTO   MELBOURNE   WELLINGTON
CAPE TOWN   IBADAN   NAIROBI   DAR ES SALAAM   LUSAKA   ADDIS ABABA
DELHI   BOMBAY   CALCUTTA   MADRAS   KARACHI   LAHORE   DACCA
KUALA LUMPUR   SINGAPORE   HONG KONG   TOKYO

541.3
H87r
90416
oct 1974

SET IN MONOPHOTO BY
D. REIDEL BOOK MANUFACTURERS, DORDRECHT, HOLLAND

PRINTED IN GREAT BRITAIN BY
J. W. ARROWSMITH LTD., BRISTOL, ENGLAND

To my wife,
for her encouragement and inspiration
in all my work

# Editor's foreword

RADIATION chemistry is not easily assigned to any one branch of chemical science, since it draws upon a wide range of disciplines and methods; in turn, studies in radiation chemistry have made new and exciting contributions in several of the more conventional areas of chemistry. Developments have now reached the stage where all chemists need to have some knowledge about the subject, and this book is to be welcomed in that it presents a concise, overall view of the present state of study.

Because radiation chemistry overlaps so extensively with other areas, many of the topics dealt with in this volume overlap with those treated in other volumes in the series, such as reaction mechanisms, magnetic resonance, and ions in solution. The books on these topics will therefore be useful as supplementary reading.

A.K.H.

# Preface

RADIATION chemistry was born out of the need to understand the mechanism of the interaction of ionizing radiations with matter, particularly in the fields of the exposure of living systems to radiation and of atomic reactor technology. In the last two decades, however, it has begun to make important contributions to other areas of chemistry. The discovery of the hydrated electron and the study of fast reactions in pulse radiolysis are sufficiently important to warrant discussion in any undergraduate chemistry course. However the pressure of other topics very often leaves time for only a fleeting discussion. This book is written to provide the reader with a short general survey of the field of radiation chemistry at present and is intended to supply the background against which these discoveries might be more properly appreciated. As such, it is hoped that it will be of interest to all undergraduates and to beginning research workers.

I would like to express my thanks to Mr. R. F. Pugh who read the manuscript and made many useful suggestions, and to Miss L. Jackson for her excellent typing of the manuscript.

G. H.

—

# Contents

# 1. Development of radiation chemistry

**Historical introduction**

RADIATION chemistry is the study of the chemical effects induced by ionizing radiations—X- and $\gamma$-rays, electrons, protons, deuterons, $\alpha$-particles, and neutrons. The earliest recorded observations in radiation chemistry were probably those of Homer who commented on the sulphurous smell in the atmosphere after lightning. This smell we now know to be due to the oxides of nitrogen and ozone produced as a consequence of reactions initiated by the lightning discharge. Since the lightning discharge is not, however, a convenient laboratory tool, radiation chemistry cannot properly be said to have begun on a scientific basis until the end of the nineteenth century when the discovery of radioactivity led to the isolation of radioactive sources.

Amongst the earliest observations of reactions induced by ionizing radiations were those of the blackening of a photographic plate exposed to uranium by H. Becquerel in 1896, the decomposition of water containing radium salts by M. Curie and A. Debierne in 1901 and the formation of ozone in oxygen irradiated with $\alpha$-particles by S. C. Lind in 1912. Little progress was made during the years prior to the Second World War and only a few people were interested in what then seemed to be a rather unimportant part of chemistry.

The production of the first nuclear reactor at Chicago during the Second World War gave a great impetus to the study of radiation chemistry. The development of the reactor emphasised the need for information on the interaction of radiation with a wide variety of materials. Thus when it was decided to cool the reactor with water circulating in aluminium pipes, there were some fears that the activation of the water by the intense radiation might lead to its reacting with aluminium as readily as it does with sodium! At the same time there was an increasing awareness of the health hazards of radiation but an understanding of the effect of radiation on living things depended on some knowledge of its interaction with relatively simple chemical systems. The synthesis of radioisotopes in the reactor also led to a greater availability of suitable radiation sources.

Some idea of the growth of radiation chemistry may be gauged from the fact that whereas during the period 1890–1920 an average of 2–3 papers a year in this subject were published, in 1970 there were about 200 publications dealing with aqueous systems alone. Far from being a rather unimportant branch of chemistry, radiation chemistry may now be said to have come of age in so far as it is making important contributions to our understanding of other areas of chemistry, as will become clear in successive chapters.

**Radiation sources**

Radiation sources are of two types—isotope sources and machine sources. Isotope sources provide continuous radiation and machine sources may provide continuous or intermittent radiation.

*Isotope sources*

Radioisotopes may be either naturally occurring or artificially produced. The characteristics of the more important radioisotopes are described in Table 1.1.

TABLE 1.1
*Radioisotopes used as radiation sources*

| Source | Half life | Radiation | Energy (MeV) |
|--------|-----------|-----------|--------------|
| $^{60}$Co | 5·27 years | $\beta$ | 0·314 |
|  |  | $\gamma$ | 1·332 |
|  |  | $\gamma$ | 1·173 |
| $^{137}$Cs | 30 years | $\beta$ | 0·52 |
|  |  | $\gamma$ | 0·662 |
| $^{90}$Sr | 28·0 years | $\beta$ | 0·544 |
| $^{32}$P | 14·22 days | $\beta$ | 1·710 |
| $^{3}$H | 12·26 years | $\beta$ | 0·018 |
| $^{222}$Rn | 3·83 days | $\alpha$ | 5·49 |
| $^{226}$Ra | 1620 years | $\alpha$ | 4·777 |
| $^{210}$Po | 138 days | $\alpha$ | 5·304 |

Most irradiation studies have been carried out using $\gamma$-emitters since very powerful sources can be obtained and dosimetry is relatively easy. Moreover because of their great penetrating power it is possible to irradiate relatively large samples in a fairly homogeneous fashion. However because of their penetrating power such sources need adequate shielding. The halfvalue thicknesses (i.e. the thickness of absorber required to reduce the intensity of the radiation by one half) for $^{60}$Co $\gamma$-rays for lead, concrete and water are respectively 1, 5 and 11 cm. A plan of a $\gamma$-irradiation facility is shown in Figure 1.1.

The source is housed in a lead pot, the front of which can be lowered by remote control and the source pushed out into the irradiation room. While the source is inside the pot, the sample to be irradiated can be positioned and, after the operator has withdrawn, the sample is irradiated. The concrete walls are 45 cm thick, and the radiation level outside the facility is safe. It is possible to use a pond facility in which the source is housed at the bottom of a tank sunk in the ground and containing water to a depth of 3 m. This depth of

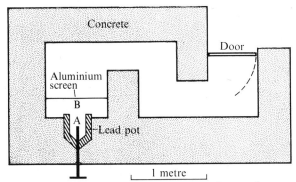

A position of retracted source   B position of exposed source

Fig. 1.1   Plan of $\gamma$-irradiation facility, Liverpool University

water will provide adequate shielding for a 2000 curie $^{60}$Co source.[†] The samples to be irradiated are then lowered to the bottom of the pond.

Most sources use $^{60}$Co which, as may be seen from Table 1.1, emits monoenergetic $\gamma$-rays of energies 1·17 and 1·33 MeV and has a half life of 5·27 years. It is prepared by neutron bombardment of $^{59}$Co in a nuclear reactor. A few sources use $^{137}$Cs which has a longer half life and consequently needs to be renewed less often. $^{137}$Cs has the advantage of being readily available as a product of the fission of uranium. The $\beta$-particles emitted from both $^{60}$Co and $^{137}$Cs are usually filtered out by the metal canisters in which the sources are contained. Typically the sources used range from activities of tens of curies in the smaller research laboratories to 650,000 curies in the Westminster Carpet Company's irradiation plant in New South Wales. The dose rates of these sources range from $10^{17}$–$10^{21}$ eV l$^{-1}$s$^{-1}$.

The most useful sources of $\beta$-particles are $^{90}$Sr, $^{32}$P and $^{3}$H. The penetrating power of $\beta$-particles is much less than that of $\gamma$-rays so that considerably less protection is required. The maximum range for $\beta$-particles from $^{90}$Sr is 0·18 cm in water and 185 cm in air. However care in handling the sources is still of the utmost importance since there is some danger of ingesting the material. $^{90}$Sr is particularly hazardous in that it is readily taken up by bone. $\beta$-sources typically range in activity from a few millicuries to 1 curie. They may be used either as external sources or in some cases as internal sources for which purpose either the source encapsulated in a thin walled vessel is placed within

---

[†] The curie (symbol Ci) is the standard unit of radioactivity and is defined as the amount of radionuclide in which the number of disintegrations per second is $3·700 \times 10^{10}$. This is very nearly the rate of disintegration of 1 g of radium.

the sample to be irradiated or a suitable compound of the radioisotope is mixed with the sample. It follows that compounds containing such radio-isotopes must necessarily be irradiated by their own $\beta$-particles and the self decomposition of compounds containing particularly $^{14}C$ and $^3H$ has been studied in some detail.

The energetics of $\beta$-particle decay are such that it is not possible to obtain monoenergetic $\beta$-particles. A typical energy spectrum for $\beta$-decay is shown in Figure 1.2.

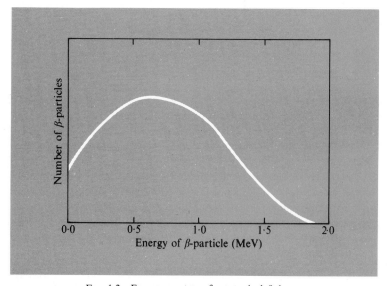

FIG. 1.2   Energy spectrum for a typical $\beta$-decay

It will be seen that although there is a clearly defined maximum energy, in practice the whole spectrum of energies from zero up to the maximum value is observed. The average energy of a $\beta$-particle is approximately one third the maximum value. The energies quoted in Table 1.1 are maximum values.

The earliest systematic studies in radiation chemistry around 1900 were primarily concerned with $\alpha$-particles since the radioelements radium, radon and polonium which had just been discovered were convenient sources of these. The $\alpha$-particles from a given emitter are monoenergetic. The range of $\alpha$-particles depends on their energy and is generally very small. For example, $\alpha$-particles from $^{210}Po$ have a range in air and water of 3·8 cm and $3·9 \times 10^{-3}$ cm respectively. For this reason, sources had either to be contained in capsules with very thin walls or else mixed with the reactants. The gas radon was a convenient source for the study of reactions in the gas phase since it could be mixed directly with the gas to be studied. However the activities that can be obtained using the naturally occurring radioelements are necessarily limited and where it is particularly desirable to study the effect of $\alpha$-particles, it is more convenient today to produce these in the cyclotron.

*Machine Sources*

The earliest machine source of radiation was the X-ray tube. It is still a convenient source of continuous irradiation particularly for studies in the energy range 0·1–0·3 MeV, as provided by a conventional medium voltage X-ray tube. X-rays are produced whenever high energy electrons are rapidly decelerated, as occurs when they pass through the electric field of an atomic nucleus. In the X-ray tube, electrons from a filament are accelerated *in vacuo* towards a water-cooled metal anode. X-rays are emitted when the electrons collide with the anode. Although medium voltage X-ray sets should be housed in a room separate from the operator, they require no special shielding and have the advantage that they can be rendered inactive by simply switching off the voltage. Higher energy X-rays can be generated using fast electron generators but these then require special shielding and do not possess any particular advantage over conventional $\gamma$-ray sources.

The cyclotron is a convenient source of $\alpha$-particle radiation. Helium ions are produced by an electric discharge in helium and, by application of a magnetic field, are constrained to move in a spiral path consisting of a series of semi-circles of gradually increasing radius. A high-frequency potential is applied, the frequency being such that the helium ion is accelerated as each semi-circle is completed. A 150 cm (60″) diameter cyclotron is capable of producing helium ions of energies up to 40 MeV with continuous beam currents of hundreds of microamperes. Because of the low penetrating power of $\alpha$-particles, samples to be irradiated must be housed in vessels with thin windows and the sample needs to be stirred vigorously so that a sufficient volume can be brought within the irradiation zone. Care needs to be taken in handling after the irradiation since metals and glass exposed to $\alpha$-particles may themselves become radioactive. By contrast, materials irradiated with $\gamma$-rays of normal energies do not become radioactive and this is another reason for the popularity of $\gamma$-ray emitters as radiation sources.

The most important source of pulsed radiation is the linear electron accelerator. Pulses of electrons are injected into a linear segmented tube down which they are accelerated by a synchronized radiofrequency wave from a klystron power valve. The r.f. field is typically $50 \text{ kV cm}^{-1}$ so that after travelling 20 cm the electron will have an energy of 1 MeV. The earliest experiments in radiation chemistry used accelerators which gave pulses of duration of 1 microsecond with repetition rates of several hundred per second. Electron energies were in the range 1–10 MeV, beam currents 0·1–2·0 A and total energy per pulse 0·2–10 J. Subsequently accelerators capable of delivering nanosecond pulses were developed and more recently picosecond pulses have been obtained.

The Van de Graaff accelerator can be used to give either pulsed or continuous irradiation. In this a moving belt transfers charge from a low-voltage supply to an insulated metal sphere which thereby builds up a high potential. This potential is then used to accelerate electrons through an evacuated tube and out through a thin window to the sample to be irradiated. Compared with the linear accelerator, the Van de Graaff accelerator is uneconomic for electron energies greater than 4 MeV.

*Pulse radiolysis*

Pulse radiolysis is to radiation chemistry what flash photolysis is to photochemistry and its application to radiation chemistry has led to discoveries of extreme importance. Although flash photolysis was developed first, the extremely short pulses that could be obtained from linear accelerators seemed at one time to make pulse radiolysis a more valuable technique. However with the development of laser light sources, the time resolution of the two methods is about the same.

A typical set up for pulse radiolysis is shown in Figure 1.3.

In *kinetic spectrophotometry*, light from a continuous light source passes through the solution being irradiated and then through a monochromator to select a given wavelength which is collected on a phototube or photomultiplier. The signal from the photomultiplier, amplified if necessary, is finally displayed on an oscilloscope where it can be photographed. By silvering the optical ends of the irradiation cell, or by a system of mirrors, it is possible to pass the beam of light through the irradiation cell several times. This gives a much enhanced effective path length and thus increases the sensitivity of the method. The high energy pulse produces a high initial concentration of reactive intermediates. The change in concentration of these intermediates can then be followed by the changes in optical density of the irradiated solution at a preselected wavelength, usually the absorption maximum. It is possible to determine the spectrum of an intermediate by measuring the optical density at several different wavelengths in a series of experiments.

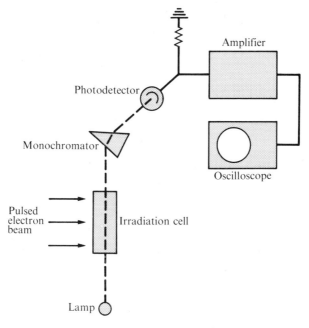

FIG. 1.3    Diagram showing basic arrangement of apparatus for pulse radiolysis

Where the complete spectrum of an intermediate is required, it is customary to use *flash spectroscopy*. In this the continuous light source is replaced by a flash lamp which is triggered by the electron pulse to give a light flash after a predetermined time interval. The light emerging from the irradiation cell is then allowed to fall on a spectrograph and the spectrum is photographed. A series of spectra can be obtained for different delay times after the pulse and thus the changes taking place can be determined. The light flash should have a lifetime similar to that of the electron pulse to give the maximum resolution of transient spectra.

Clearly only those intermediates with lifetimes comparable to or greater than the electron pulse lifetime will be detected by either method. Thus the majority of experiments carried out to date in pulse radiolysis have been concerned with intermediates having lifetimes of microseconds or longer although some experiments have been carried out on intermediates of nanosecond lifetimes. Results are just beginning to emerge from picosecond radiolysis and the technique should continue to yield valuable information for the next few years.

## PROBLEMS

1.1. The equation of radioactive decay is

$$\frac{dN}{dt} = -\lambda N$$

where $N$ is the number of radioactive atoms present at time $t$ and $\lambda$ is the disintegration constant for the radioisotope. This integrates to give the equation

$$N = N_0 e^{-\lambda t}$$

where $N_0$ is the number of atoms present at zero time. Show that $\lambda$ is related to half life of the radioisotope by

$$t_{1/2} = \frac{\log_e 2}{\lambda}$$

What weight of pure $^{137}CsCl$ would be required to give an activity of 1000 Ci? See pp. 2 and 3.

1.2. Calculate the activity of a 500 Ci $^{60}Co$ source after (a) 1 year (b) 10 years. Repeat the calculation for a 500 Ci $^{137}Cs$ source and comment on the significance of your results.

1.3. Calculate the number of $\beta$-particles emitted per year in 10 g water containing 5 percent $^{3}H_2O$. The equation for radioactive decay is

$$^{3}_{1}H \rightarrow {}^{3}_{2}He + {}^{0}_{-1}e$$

1.4. 0·1 cm$^3$ $^{222}Rn$ was mixed with 1 litre of acetylene (volumes measured at N.T.P.). Calculate the number of $\alpha$-particles emitted and hence the energy absorbed by the acetylene per day.
The equation for radioactive decay is

$$^{222}_{86}Rn \rightarrow {}^{218}_{84}Po + {}^{4}_{2}He$$

(To a first approximation, all $\alpha$-particles will be absorbed by the acetylene).

# 2. Interaction of radiation with matter

## Energy loss of γ-rays

GAMMA-RAYS are electromagnetic radiation of very short wavelength, about 0·01 Å (0·001 nm) for a 1 MeV photon. Absorption of γ-rays by matter obeys the fundamental Lambert–Beer law

$$I = I_0 e^{-\mu x}$$

where $I$, $I_0$ are the intensities of the transmitted and incident radiations respectively, $x$ is the thickness of the absorber and $\mu$ is the linear absorption coefficient. If the thickness of the absorber is expressed in cm, $\mu$ will have the units cm$^{-1}$. The linear absorption coefficient depends on the density of the absorber and it is convenient to define the mass absorption coefficient by

$$\mu_{mass} = \mu_{linear}/\rho$$

where $\rho$ is the density of the material. The mass absorption coefficient (units cm$^2$g$^{-1}$) is the same for the different physical states, i.e. gaseous, liquid and solid, of the absorber.

The total absorption coefficient is the sum of three separate coefficients representing the three main processes of energy absorption by γ-rays. These processes are the photoelectric effect, the Compton effect, and pair production.

### Photoelectric effect

In the photoelectric effect, the entire energy of the γ-ray photon is transferred to an atomic electron with subsequent ejection of the electron, usually from the K-shell. The electron then proceeds in the medium with an energy equal to that of the original quantum less the binding energy of the electron in the atom. Binding energies range from 100 eV for low atomic weight materials to 100 KeV for high atomic weight materials. The inner orbital vacancy is subsequently filled by an outer electron of the atom with consequent liberation of energy. This energy may appear as characteristic X-radiation or may result in the emission of further electrons—the Auger effect. The photoelectric effect is greatest for low photon energies, <0·1 MeV, and for elements of high atomic number.

### Compton effect

In the Compton effect, the γ-ray photon gives up only part of its energy to an electron, which may be free or bound in an atom. In the latter case the electron will be ejected from the atom. The incident photon is scattered and proceeds with an energy equal to that of the original quantum less the recoil energy of the electron. The scattered photon may then undergo subsequent absorption either by the Compton effect or the photoelectric effect. The Compton effect depends only on the number of electrons per gram of material. For an element of atomic number $Z$ and atomic weight $A$ the dose absorbed per unit mass of material will be proportional to $Z/A$. For a compound the ratio $\Sigma Z/\Sigma A$ is used, e.g. for $H_2O$

$$\Sigma Z/\Sigma A = (2+8)/(2+16) = 10/18$$

If the absorbed dose is expressed per unit volume of material then it will be proportional to the electron density i.e. number of electrons per unit volume of material. It is usually convenient to evaluate this as the number of gram electrons per $cm^3$ of sample.

Compton absorption is more important for photons of higher energies and is, for example, the only absorption process for the interaction of $^{60}Co$ $\gamma$-rays with water.

*Pair production*

In pair production the $\gamma$-ray photon is completely absorbed within the vicinity of an atomic nucleus and produces a positron–electron pair. The energy of the photon must be at least $2\ m_ec^2$, i.e. $1.02$ MeV, where $m_e$ is the rest mass of the electron. Any excess energy of the photon appears as kinetic energy of the electron and positron. The positron and electron, after being slowed down, recombine with each other resulting in the production of two $0.51$ MeV $\gamma$-rays (annihilation radiation).

The dependence of the relative importance of the three absorption processes on the energy of the incident $\gamma$-radiation may be seen from Figure 2.1.

Whatever the mechanism of energy loss of the $\gamma$-ray photon, secondary electrons with considerable kinetic energy are produced. Subsequent energy absorption processes, which will account for most of the energy of the incident photon, will be those characteristic of electrons and will therefore be similar to those occurring when the material is bombarded directly with electrons. The processes of energy loss by electrons have been investigated in some detail.

**Energy loss of electrons**

Electrons interact with matter in several ways, chief of which are the emission of electromagnetic radiation and inelastic and elastic collisions.

*Emission of radiation*

An electron passing close to the nucleus of an atom will be decelerated

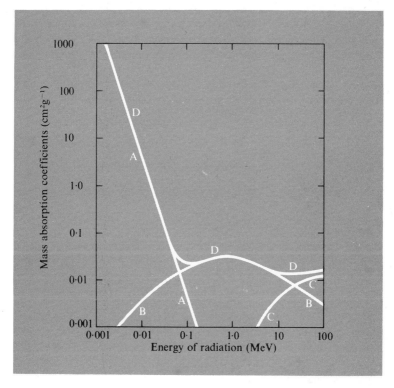

FIG. 2.1   Dependence of the mass absorption coefficients of water on energy of
γ-radiation
A  absorption by photoelectric effect
B  absorption by Compton effect
C  absorption by pair production
D  total absorption coefficient

with the consequent emission of energy as an X-ray photon. This radiation is
known as *bremsstrahlung* and though its production is not accompanied by
any chemical change in the medium, it may be subsequently absorbed and
lead to chemical change. *Bremsstrahlung* emission is the predominant mode
of energy loss for electrons of energies in the range 10–100 MeV but is negligible
below 0·1 MeV. Cerenkov radiation is emitted as a characteristic blue glow
when high-energy electrons are slowed down to the speed of light in materials
of refractive index greater than one.

*Inelastic collisions of electrons*

At lower energies electrons lose their energy by inelastic collisions with electrons of the stopping material resulting in ionization and excitation in the material. The equation describing the rate of energy loss of an electron by ionization and excitation was first derived by H. A. Bethe and is given by

$$\frac{\mathrm{d}E}{\mathrm{d}x} = \frac{2\pi Ne^4 Z}{m_0 v^2} \left[ \ln \frac{m_0 v^2 E}{2I^2 (1-\beta^2)} - \ln 2 \{2\sqrt{(1-\beta^2)} - 1 + \beta^2\} \right.$$
$$\left. + 1 - \beta^2 + \tfrac{1}{8}\{1 - \sqrt{(1-\beta^2)}\}^2 \right] \qquad (2.1)$$

where $E$ is the energy of the electron
$x$ is the distance travelled by the electron in the material
$N$ is the number of atoms per $\mathrm{cm}^3$
$e$ is the charge on the electron
$Z$ is the atomic number of the stopping material
$m_0$ is the rest mass of the electron
$v$ is the velocity of the electron
$I$ is the mean excitation potential for the atoms of the stopping material
and    $\beta = v/c$ where $c$ is the velocity of light.

The quantity $\mathrm{d}E/\mathrm{d}x$, i.e. the rate of loss of energy with distance, is usually known as the linear energy transfer (symbol LET) and is a very important quantity in describing the behaviour of any radiation. The main conclusions to be drawn from equation (2.1) about the LET of electrons are:

(1) It increases as the energy decreases, i.e. as the electrons are slowed down in the medium.

(2) The product $NZ$ is equal to the number of electrons per $\mathrm{cm}^3$, i.e. LET is directly proportional to electron density and will therefore be much lower in gases than in solids or liquids.

*Elastic collisions of electrons*

Electrons because of their small mass are readily deflected by the Coulombic field of a nucleus. Although there is no energy loss in elastic collisions of this kind, such collisions are important in that they result in a non-linear passage of the electron through the medium. Elastic collisions are important for low-energy electrons in materials of high atomic number.

**Energy loss of other charged particles**

The interaction of other charged particles, e.g. α-particles, protons and deuterons with matter is similar to that of electrons. The main mechanism of energy loss is by inelastic collisions with the electrons of the stopping material and equations for the rates of energy loss of such particles have been worked out.

**Effect of LET**

The LET of all particles increases as the energy of the particle decreases. It is, however, convenient to characterize radiations by their mean LET values, i.e. the total energy of the radiation divided by the distance travelled by the radiation before coming to rest. Mean LET values for different radiations are given in Table 2.1. Usually LET values are in units of eV per ångstrom. The conversion to SI units is discussed on p. 17.

TABLE 2.1
*Mean LET values for water*

| Radiation | Mean LET (eV Å$^{-1}$) |
|---|---|
| 2 MeV electrons | 0·021 |
| 1.25 MeV $\gamma$-rays ($^{60}$Co) | 0·021 |
| 0.018 MeV $\beta$-particles ($^3$H) | 0·32 |
| 5.3 MeV $\alpha$-particles ($^{210}$Po) | 13·6 |
| 1 MeV $\alpha$-particles | 18·9 |

It is instructive to follow the passage of a high-energy electron ($\sim$2MeV) through a typical liquid, e.g. water. Secondary electrons will be produced along the primary track as a result of inelastic scattering. Some of these electrons will have relatively low energies, about 100 eV, and this energy will be dissipated rapidly as ionization and excitation within a relatively short range, about 20 Å. Such ionizations and excitations are said to lie within a spur and for electrons of LET 0·02 eV Å$^{-1}$, such spurs will lie on average 5000 Å apart. The energy required to create an ion pair is about 30 eV and about three ion pairs will be produced within a spur. Other secondary electrons will have relatively high energies with characteristic low LET and will form tracks of their own branching off from the primary track. Such tracks are known as $\delta$-rays. Towards the end of a track when the energy of the electron falls below 500 eV, the LET will be much higher and the region of ionization both bigger and denser. These are sometimes referred to as blobs. With more densely ionizing radiation, the spurs will be produced in such proximity that they overlap each other resulting in the production of a high concentration of ions and excited molecules along the primary track.

Typical patterns of spur production are shown in Figure 2.2.

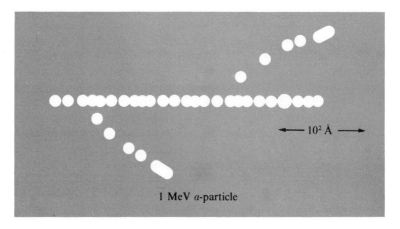

$10^2$ Å

1 MeV $\alpha$-particle

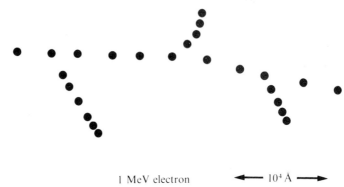

1 MeV electron        $10^4$ Å

FIG. 2.2  Production of spurs along tracks of ionizing particles. Spurs are approximately 20 Å in diameter and are not drawn to scale in the case of the 1 MeV electron

The initial distribution of reactive species is very inhomogeneous. The entities produced within a spur may recombine to give molecular products or they may diffuse out into the bulk of the solution where they may react with other solutes. The yields of species recombining and diffusing out are known as the molecular and radical yields respectively.

The pattern of energy loss in solids will be similar to that in liquids but in gases the LET is considerably less because of the lower density. Consequently even for $\alpha$-particles the overlap of spurs is much less significant and radical rather than molecular products predominate.

### Time scale of events

Table 2.2 shows the time scale of the principal events occurring in radiolysis. These times depend to a small extent on the particular substrate and are therefore approximate.

TABLE 2.2

*Time scale of events in radiation chemistry*

| Time (s) | Event |
|---|---|
| $10^{-18}$ | Ionizing radiation traverses one molecule. |
| $10^{-15}$ | Time interval between successive ionizations. |
| $10^{-14}$ | Dissociation of electronically excited species. Transfer of energy to vibrational modes. Ion–molecule reactions beginning. |
| $10^{-13}$ | Electrons reduced to thermal energies. |
| $10^{-12}$ | Radicals diffuse one jump. |
| $10^{-11}$ | Electron is solvated in polar media. |
| $10^{-10}$ | Fastest diffusion controlled reactions are complete. |
| $10^{-8}$ | Molecular products complete. Radiative decay of excited singlet states. |
| $10^{-5}$ | Scavenging of radicals by reactive scavengers. |
| $10^{-3}$ | Radiative decay of triplet states. |
| 1 | Most reactions are complete. In some systems, post-irradiation reactions may continue for several days. |

It will be seen that, apart from the earliest events, many of these processes can be studied using the technique of pulse radiolysis. In the solid state, some of the reactive intermediates are trapped and can be analysed by conventional spectroscopy.

### Comparison with photochemistry

It is illuminating to compare radiation chemistry with photochemistry. In photochemistry the radiation will usually be monochromatic and one quantum of the radiation will be entirely absorbed by one molecule leading to the production, at least initially, of only one excited state. The nature of the excited state will be dictated by the optically allowed transitions. The absorption of radiation will generally be highly specific so that, even in a dilute solution, all the radiation may be absorbed by the solute. The excited states produced will be distributed homogeneously in the solution.

In radiation chemistry the radiation may or may not be monochromatic, but even if it is initially monochromatic, the processes of energy loss are such that it will pass through a spectrum of lower energy states. In so doing, the initial quantum or particle will bring about the excitation and ionization of many molecules. Excitation of molecules by slow electrons is not governed by the optical selection rules, and states that are not readily observed in photolysis, e.g. triplet states, may be readily produced. The absorption of radiation will be non-specific and in a solution each component will absorb energy in proportion to its electron fraction†. In a dilute solution most of the energy is absorbed by the solvent. The reactive species produced are not distributed homogeneously in the solution.

Photochemical systems are sometimes difficult enough to understand. In view of the much greater complexity of radiation chemical systems, it must be wondered whether the study of such systems is a case of fools rushing in where angels fear to tread. It is gratifying to report, however, that the results obtained from radiation chemistry have been of value not only to photochemistry but also to other branches of chemistry, as will become evident in succeeding chapters.

### Dosimetry

*Units of dose*

The International Commission on Radiological Units and Measurement in 1959 recommended the adoption of the rad as the unit of absorbed dose of any ionizing radiation. 1 rad is 100 ergs per gram. Although this unit is commonly in use amongst medical physicists and radiobiologists, chemists have found it more convenient to measure absorbed doses in units of $eV\ g^{-1}$ or $eV\ cm^{-3}$. Radiation yields are then defined as $G$ values, where $G(X)$ is the number of atoms or molecules of a species X produced per 100 eV energy absorbed.

The intensity of X- or $\gamma$-radiation is sometimes expressed in roentgens (symbol R). One roentgen is the quantity of X- or $\gamma$-radiation such that the associated corpuscular emission per 0·001293 g of air (the weight of 1 cm³ at N.T.P.) produces, in air, ions carrying 1 electrostatic unit (e.s.u.) of electricity of

---

† electron fraction of component A $= \dfrac{\text{number of electrons in A}}{\text{total number of electrons in system}}$

either sign. Since the electronic charge is $4\cdot80298 \times 10^{-10}$ e.s.u. this means that 1 R produces $1\cdot61 \times 10^{12}$ ion pairs per gram of air. This corresponds to the absorption of $87\cdot7$ ergs per gram of air. The roentgen is an exposure dose rather than an absorbed dose and the actual dose absorbed by the sample will be determined by the number of electrons per unit weight of sample. Thus the dose absorbed by air and water when exposed to 1 R are $5\cdot48 \times 10^{13}$ eV g$^{-1}$ ($0\cdot88$ rad) and $6\cdot06 \times 10^{13}$ eV g$^{-1}$ ($0\cdot973$ rad). Within a few per cent 1 R of radiation will result in the absorption of 1 rad by water or living tissue composed largely of water.

Although the eV is not an SI unit there is as yet little tendency amongst radiation chemists to abandon this unit. Appropriate conversion factors together with other useful radiological information are summarised in Table 2.3.

TABLE 2.3
*Conversion factors and other data*

$$
\begin{aligned}
1 \text{ rad} &= 100 \text{ erg g}^{-1} \\
&= 6\cdot24 \times 10^{13} \text{ eV g}^{-1} \\
&= 10^{-5} \text{ J g}^{-1} \\
&= 2\cdot40 \times 10^{-6} \text{ cal g}^{-1} \\
1 \text{ e.s.u.} &= 3\cdot334 \times 10^{-10} \text{ C} \\
1 \text{ eV} &= 1\cdot6021 \times 10^{-19} \text{ J} \\
1 \text{ eV per molecule} &= 96\cdot48 \text{ kJ mol}^{-1} \\
&= 23\cdot06 \text{ kcal mol}^{-1} \\
\text{LET of } 1 \text{ eV Å}^{-1} &= 1\cdot602 \text{ aJ nm}^{-1} (\text{aJ} = 10^{-18} \text{ J}) \\
1 \text{ Ci} &= 3\cdot700 \times 10^{10} \text{ disintegrations per second.}
\end{aligned}
$$

1 Ci of $^{60}$Co gives an exposure rate of $1\cdot32$ *R* per hour at 1 m. The resultant absorbed dose rates in air and water are $1\cdot15$ and $1\cdot27$ rad per hour.

From the data of Table 2.3 it follows that a yield of 1 molecule per 100 ev is equal to

$$\frac{1}{100} \times \frac{1}{6\cdot02 \times 10^{23}} \times \frac{1}{1\cdot6021 \times 10^{-19}} \text{ mol J}^{-1} = 1\cdot04 \times 10^{-7} \text{ mol J}^{-1}$$

Thus if radiation yields were redefined as the number of moles per $10^7$ J, within a few per cent they would have the same values as on the electron-volt standard.

*Techniques of measurement*

The experimental measurements of absorbed dose fall into two groups—absolute and secondary. An absolute measurement of dose can be made using an ionization chamber. Basically this consists of two electrodes separated by a space containing a suitable gas which is ionized by the incident radiation. Ions produced in the gas are collected at the electrodes by means of an applied potential and the number of ions collected can be counted. Provided the energy required to create an ion pair in the gas is known, the absorbed dose can be calculated. Alternatively the beam of X- or $\gamma$-rays can be absorbed in a block of material and the resultant temperature rise measured. Clearly the material needs to be such that none of the energy goes into chemical reaction. Moreover it should have high thermal conductivity so that the temperature is uniform. Metals and graphite satisfy these criteria.

Since absolute measurements of dose involve very careful techniques, they are not suitable for routine measurements where ease and speed are essential. For this reason secondary methods are usually employed. The most common dosimeter in use in radiation chemistry laboratories is the Fricke dosimeter. This consists of an air-saturated aqueous solution of $10^{-3}$ M iron (II) ammonium sulphate, $10^{-3}$ M sodium chloride and 0·4 M sulphuric acid. Triply distilled water and A. R. purity reagents are used in making up the solution. It is known by comparison with calorimetric measurements that $G$ (Fe (III)) for this solution when irradiated with $^{60}$Co $\gamma$-rays is 15·6. The reactions occurring in this system are discussed in Chapter 6. The ferric ion is conveniently determined from the optical density at 303 nm, where its extinction coefficient is $2·20 \times 10^3$ M$^{-1}$ cm$^{-1}$ at 298 K. The sample to be irradiated is then placed in an identical position to the dosimeter. For $^{60}$Co $\gamma$-rays the absorption is entirely by the Compton process and the energy absorbed is proportional to the electron density of the sample.

CALCULATION: The optical density of Fricke dosimeter solution after 10 minutes irradiation with $^{137}$Cs $\gamma$-rays was 0·440 at 303 nm at 298 K in 1 cm cells. Calculate
a) the dose rate given to the solution, and
b) the dose rate for liquid methanol in an identical position.

The optical density, O.D., of a solution containing a concentration, $c$, of an absorber of extinction coefficient, $\varepsilon$, in a path length, $l$, is given by

O.D. $= \varepsilon c l$

$\therefore$ $[\text{Fe (III)}] = \dfrac{0·440}{2·20 \times 10^3} = 2 \times 10^{-4}$ M

$\therefore$ No. of ions of Fe(III) $= 6·02 \times 10^{23} \times 2 \times 10^{-4}$ $l^{-1}$
but 15·6 ions Fe(III) are produced per 100 eV

$\therefore$ Dose rate $= \dfrac{6·02 \times 10^{23} \times 2 \times 10^{-4}}{15·6} \times 100 \times \dfrac{1}{10 \times 60}$ eV $l^{-1}$s$^{-1}$

$= 1·28 \times 10^{18}$ eV $l^{-1}$s$^{-1}$

Energy absorption of $^{137}$Cs $\gamma$-rays will be entirely by Compton process and is therefore proportional to electron density.

Density of 0·4 M $H_2SO_4$ is 1·024 g cm$^{-3}$ at 293 K.
1 litre of 0·4 M $H_2SO_4$ contains $1024 - 0·4 \times 98$ g $H_2O$
1 molecule of $H_2SO_4$ contains 50 electrons
1 molecule of $H_2O$ contains 10 electrons

$$\therefore \quad \text{electron density} = \frac{0·4 \times 50 + 985/18 \times 10}{1000} \text{ g electrons cm}^{-3}$$
$$= 0·567 \text{ g electrons cm}^{-3}$$

Density of methanol is 0·791 g cm$^{-3}$ at 293 K.
1 molecule of $CH_3OH$ contains 18 electrons

$$\therefore \quad \text{electron density is } \frac{18}{32/0·791} \text{ g electrons cm}^{-3}$$
$$= 0·445$$

The dose rate in methanol will be $1·28 \times 10^{18} \times \dfrac{0·445}{0·567} = 1·01 \times 10^{18}$ eV $l^{-1}$s$^{-1}$

It will be seen that the ratio

$$\frac{\text{electron density in methanol}}{\text{electron density in 0·4 M } H_2SO_4} = \frac{0·445}{0·567} = 0·785$$

This is very nearly the same as the ratio of densities, and in fairly similar liquids, to a first approximation, the dose absorbed may be assumed to be simply proportional to density.

Other chemical dosimeters have been described but generally are not so convenient. $N_2O$ is extensively used in the dosimetry of gases. Where only approximate measurements of dose are required, solid-state dosimeters may be used. The colour produced in silver-activated phosphate glass is a measure of dose. Workers in radiation chemistry usually wear a film badge and the fogging produced in the photographic film is a measure of the dose received by the operator.

## PROBLEMS

2.1. Calculate the percentage of energy absorbed by the benzene in a solution of 10 g of benzene in 100 g of acetone when the solution is irradiated with
a) ultraviolet light of wavelength 260 nm.
b) 1 MeV electrons.
The extinction coefficients of benzene and acetone at 260 nm are 170 and 18 M$^{-1}$cm$^{-1}$ respectively.
The densities of benzene and acetone are 0·879 and 0·791 g cm$^{-3}$ respectively at 293 K. (Assume no volume change on mixing).
Comment on the significance of your results.

2.2. Given the half-value thicknesses for lead and water for $^{60}$Co $\gamma$-rays are 0·6 and 11 cm respectively, calculate the mass absorption coefficients of these materials. Calculate the thickness of each of these materials that would be required to

reduce the intensity of a 1000 Ci $^{60}$Co source by a factor of $10^4$.
Density of lead is $11 \cdot 3$ g cm$^{-3}$ at 293 K.
Density of water is $0 \cdot 998$ g cm$^{-3}$ at 293 K.

2.3. Calculate the temperature rise in a block of lead when uniformly irradiated with 1 MeV $\gamma$-rays at a dose rate of 10 rad s$^{-1}$ for 10 minutes. (Assume no heat loss to the surroundings).
Specific heat of lead is $0 \cdot 126$ J K$^{-1}$g$^{-1}$.

2.4. The dose rate of a 1 MeV electron source was $10^{19}$ eV $l^{-1}$s$^{-1}$ as determined by the Fricke dosimeter. Calculate the corresponding dose rate in cyclohexane. Density of cyclohexane is $0 \cdot 779$ g cm$^{-3}$.

2.5. The range of $4 \cdot 0$ MeV $\alpha$-particles in air at 288 K and 760 mm Hg is $2 \cdot 30$ cm. Calculate the mean LET value and compare this with the results in Table 2.1. Comment on the comparison.

# 3. Ions and electrons

## Radiolysis of gases

*Evidence for the existence of charged particles*

IN Chapter 2 it was shown that energy loss by X- and $\gamma$-rays and electrons resulted in the production of electrons within the medium.

$$M \; -\!\!\!\sim\!\!\!\rightarrow \; M^+ + e^- \tag{3.1}$$

(In radiation chemistry the symbol $-\!\!\sim\!\!\rightarrow$ is used to denote 'yields under the influence of ionizing radiation'.). In the gaseous phase these electrons have characteristically low LET values and will travel some distance from their parent ion before they lose their energy, when they are said to be thermalized. They will then be captured by positive ions

$$M^+ + e^- \; \rightarrow \; M^* \tag{3.2}$$

or react with neutral molecules. Electron capture by a neutral molecule may be of the type

$$M + e^- \; \rightarrow \; M^- \tag{3.3}$$

or may lead to dissociation.

$$AB + e^- \; \rightarrow \; A + B^- \tag{3.4}$$

Much of the early work in radiation chemistry was carried out using gaseous systems. This was due to the fact that the earliest radiation sources were $\alpha$-particle emitters. Radon, being a gas, could be readily mixed with other gases ensuring uniform irradiation. The $\alpha$-particles emitted had only a short range and the energy deposited in the gas could therefore be readily calculated. As a result of the pioneering work of S. C. Lind it became clear that ions and electrons were important intermediates in many gas-phase reactions. Their existence could be directly demonstrated by enclosing the irradiated gas between two plates maintained at a potential difference. In the absence of the radiation no current flowed, but a current which increased to a saturation value at high voltages, was observed in the irradiated gas. A typical result is shown in Figure 3.1.

From the saturation current the number of ions could be calculated. Radiation yields were expressed as $M/N$ values where

$$M/N = \frac{\text{no. of molecules of gas reacted}}{\text{no. of ions formed in the gas}}$$

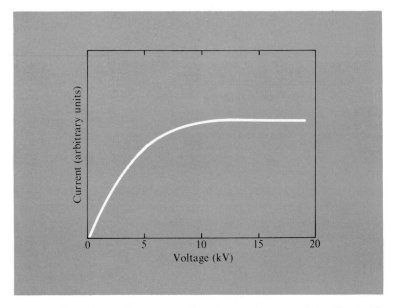

FIG. 3.1   Dependence of current on voltage for gaseous ammonia irradiated with α-particles

For most gases $M/N$ values were about one, although in a few cases, e.g. the para-ortho hydrogen conversion and the decomposition and synthesis of hydrogen bromide, much higher values were observed. It was suggested that in these systems the energy of charge neutralization was shared amongst neutral molecules which clustered around the ions and led to decomposition of these molecules. We now know that the higher yields in these reactions arise from free radical chain reactions, the free radicals being produced by decomposition of the ions. Clustering may still be important in some gas-phase reactions.

$M/N$ values can be related to $G$ values if $W$, the energy required to create an ion pair in the gas by the radiation, is known. $W$ can be determined if a known number of α-particles of known initial energy are entirely absorbed within an ionization chamber and the total ionization per α-particle measured. Values of $W$ are usually about 30 eV per ion pair and are almost independent of the LET of the radiation. $M/N$ values should be multiplied by three to give a roughly comparable $G$ value.

Some typical values of $W$ together with the corresponding ionization potentials are given in Table 3.1.

TABLE 3.1

*Energies of ion-pair production*

| Gas | $W$ (eV) | $I$ (eV) |
|---|---|---|
| $H_2$ | 36·3 | 15·4 |
| He | 42·3 | 24·6 |
| $N_2$ | 34·9 | 15·6 |
| $O_2$ | 30·8 | 12·2 |
| $CH_4$ | 27·3 | 13·0 |
| $C_2H_6$ | 24·5 | 11·7 |

It is clear from Table 3.1 that $W$ is always greater than the ionization potential and that $I \approx 0.4\ W$. Thus not all the energy of the radiation goes into producing ions and some must result in the production of excited states.

It is possible to separate the contributions of ionic and excited-state reactions. In the α-radiolysis of gaseous ammonia, $M/N$ is reduced from 1·37 in the absence of an applied field to 0·97 in a field sufficient to produce saturation. If it is assumed that at saturation all ions are discharged at the electrodes without reaction, then the results indicate that ionic processes contribute only to the extent of about 30 per cent of the overall decomposition.

*Ionic reactions in the mass spectrometer*

In the mass spectrometer, ions are produced as a result of the collision of low-energy electrons ($<100$ eV) with molecules. Some correlation might therefore be expected between the species observed in the mass spectrometer and the radiation chemistry of the gaseous material. There are however important differences between the two systems. The mass spectrometer usually operates at very low pressures ($10^{-7}$–$10^{-5}$ mm Hg) although special high-pressure mass spectrometers have been designed for the study of ion—molecule reactions. In the mass spectrometer an ion will be at some distance from a neutral molecule and will probably undergo unimolecular fragmentation. However in radiolysis, as well as ion fragmentation, ion–molecule reactions are probable.

The ions formed in the mass spectrum of methane and their relative abundances are $CH_4^+$ 47 percent, $CH_3^+$ 40 percent, $CH_2^+$ 8 percent, $CH^+$ 4 percent and $C^+$ 1 percent. The predominant species are $CH_4^+$ and $CH_3^+$ and these undergo ion–molecule reactions as follows:

$$CH_4^+ + CH_4 \rightarrow CH_5^+ + \cdot CH_3 \qquad (3.5)$$

$$CH_3^+ + CH_4 \rightarrow C_2H_5^+ + H_2 \qquad (3.6)$$

On the basis of these reactions and making some simplifying assumptions regarding the fates of other ions and radicals, it is possible to predict the radiolysis yields. The calculated and observed values are shown in Table 3.2.

TABLE 3.2

*Product yields in the 2 MeV electron irradiation of $CH_4$*

| Product | $(M/N)$ observed | $(M/N)$ calculated |
|---------|------------------|--------------------|
| $CH_4$ | $-2 \cdot 4$ | $-2 \cdot 5$ |
| $H_2$ | $1 \cdot 6$ | $1 \cdot 9$ |
| $C_2H_6$ | $0 \cdot 6$ | $0 \cdot 7$ |
| $C_3H_8$ | $0 \cdot 08$ | $0 \cdot 05$ |
| $C_4H_{10}$ | $0 \cdot 01$ | $0 \cdot 01$ |

The agreement is surprisingly good considering that the conditions varied and that the treatment takes no account of any effects due to excited molecules. It is unlikely, however, that mass spectrometric data can be extrapolated to account for the radiation chemistry of condensed systems.

**Radiolysis of liquids**

*Ionic species in the radiolysis of water and polar liquids*

An electron will lose its energy much more rapidly in condensed phases than in the gaseous phase. Before the late 1950s, two conflicting views on water radiolysis were held. The first supposed that an electron ejected from a molecule would be thermalized before it could escape the Coulombic attraction of the parent positive ion and would thus be recaptured to give an excited molecule which subsequently decomposed.

$$H_2O \xrightarrow{\hspace{0.8cm}} H_2O^+ + e^- \qquad (3.7)$$
$$e^- + H_2O^+ \rightarrow H_2O^* \qquad (3.8)$$
$$H_2O^* \rightarrow H\cdot + \cdot OH \qquad (3.9)$$

Alternatively it was suggested that the electron would escape the parent ion but would react rapidly with a neutral water molecule.

$$H_2O \xrightarrow{\hspace{0.8cm}} H_2O^+ + e^- \qquad (3.7)$$
$$e^- + H_2O \rightarrow H\cdot + OH^- \qquad (3.10)$$
$$H_2O^+ + H_2O \rightarrow H_3O^+ + \cdot OH \qquad (3.11)$$

(A positive ion in a condensed phase, e.g. $H_2O^+$, is sometimes referred to as the positive hole). By either mechanism the primary products were hydrogen atoms and hydroxyl radicals. In 1953, R. L. Platzman had suggested that the hydrated electron might exist sufficiently long to react with other solutes though there was little experimental evidence for this. He predicted that the species should have a blue colour and a hydration energy of about 2 eV.

Subsequently it was shown that there are at least two reducing species in the radiolysis of liquid water and these were tentatively identified as the hydrogen atom and the hydrated electron, $e_{aq}^-$. That the principal species in the radiolysis of neutral water had indeed unit negative charge was strikingly verified by the study of the kinetic salt effect on its reactions. The dependence of the rate coefficient on ionic strength for a reaction between ions in water at 298 K is given by the Brønsted–Bjerrum equation:

$$\log k/k_0 = 1 \cdot 02 \, Z_A Z_B \, \frac{\mu^{1/2}}{1+\mu^{1/2}} \tag{3.12}$$

where $k$, $k_0$ are the rate coefficients at ionic strengths $\mu$ and 0 respectively and $Z_A$, $Z_B$ are the algebraic numbers of the charges on the ions. The reaction of the reducing species with $H_2O_2$, $NO_2^-$, $O_2$ and $H_3O^+$ was investigated.

$$e_{aq}^- + H_2O_2 \xrightarrow{k_a} \cdot OH + OH^- \tag{3.13}$$

$$e_{aq}^- + NO_2^- \xrightarrow{k_b} NO_2^{2-} \cdot \tag{3.14}$$

$$e_{aq}^- + O_2 \xrightarrow{k_c} O_2^- \cdot \tag{3.15}$$

$$e_{aq}^- + H_3O^+ \xrightarrow{k_d} H\cdot + H_2O \tag{3.16}$$

The effect of ionic strength on the rate coefficient ratios $k_b/k_a$, $k_c/k_a$ and $k_d/k_a$ was determined and the results are shown in Figure 3.2. $K$, $K_0$ are the rate coefficient ratios at ionic strengths $\mu$ and 0 respectively. In all cases the results were in good quantitative agreement for a reductant having unit negative charge.

Direct evidence for the existence of the hydrated electron was finally obtained in 1963 when J. W. Boag and E. J. Hart, and J. P. Keene independently observed its transient spectrum in the pulse radiolysis of water. The spectrum is shown in Figure 3.3. The absorption peak at 700 nm results in the hydrated electron having a blue colour and is in striking agreement with the earlier theoretical predictions of R. L. Platzman. The allocation of this spectrum to the hydrated electron was based on the following evidence:

1. It was similar to the spectrum of the solvated electron observed in solutions of the alkali metals in liquid ammonia.
2. Typical electron scavengers—e.g. $H_3O^+$ or $N_2O$—suppressed the spectrum.

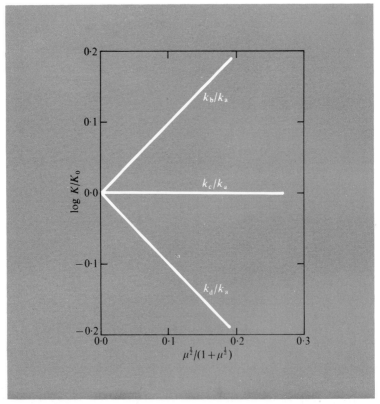

FIG. 3.2  Effect of ionic strength on rate coefficient ratios for reactions of the hydrated electron

3. The effect of ionic strength on the rate of reaction of this intermediate with $Fe(CN)_6^{3-}$ was consistent with a species of unit negative charge.

$$e_{aq}^- + Fe(CN)_6^{3-} \rightarrow Fe(CN)_6^{4-} \qquad (3.17)$$

The molar extinction coefficient was determined using tetranitromethane as scavenger. The reaction is

$$e_{aq}^- + C(NO_2)_4 \rightarrow C(NO_2)_3^- + NO_2 \qquad (3.18)$$

The nitroform anion has been well characterized in other systems and at 366 nm its extinction coefficient is $1 \cdot 02 \times 10^4 \ M^{-1} \ cm^{-1}$. It was possible in the same solution to measure the production of nitroform anion from the increase in optical density at 366 nm and the corresponding disappearance of the hydrated

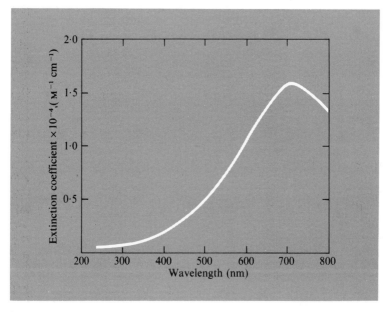

FIG. 3.3    Absorption spectrum of the hydrated electron at room temperature

electron from its optical density at 578 nm and hence to establish an absolute extinction coefficient for the hydrated electron. At its absorption maximum of 720 nm, $\varepsilon = 1{\cdot}58 \times 10^4$ $M^{-1}$ $cm^{-1}$.

The low values of the hydration energy, about 170 J $mol^{-1}$, and of the diffusion coefficient, $4{\cdot}25 \times 10^{-5}$ $cm^2$ $s^{-1}$, together with the relatively large reaction radius, about 3 Å, indicate that the electron is not attached to one water molecule only but rather that it exists in a trap formed by the polarization which it induces in the dipolar water molecules which surround it. By measuring the efficiency of a known amount of scavenger in suppressing the transient absorption, it is possible to determine the rate coefficient for the reaction of the hydrated electron with the scavenger. Reactants are restricted to those which do not absorb significantly in some region of the hydrated electron spectrum. Once the experimental facility for pulse radiolysis has been established, data can be obtained in a relatively short time. Computer pro-grammes have been developed for kinetic analysis of the oscilloscope traces and it is thus possible to obtain rate coefficients by this technique with consid-erable ease. In this way the rate coefficients for the reactions of the hydrated electron with a very wide range of reactants have been determined and it is probably true to say that the kinetic behaviour of the hydrated electron has been more widely characterized than that of any other chemical intermediate. Some typical rate coefficients are given in Table 3.3.

TABLE 3.3

*Rate coefficients of reactions of the hydrated electron*

| Reactant | pH | rate coefficient ($\mathrm{M^{-1}s^{-1}}$) |
|---|---|---|
| $e_{aq}^-$ | 10–13 | $5 \times 10^9$ |
| H | 10·9 | $3 \times 10^{10}$ |
| $H_2O$ | 8–9 | $1·6 \times 10^1$ |
| $H_2O_2$ | 7 | $1·3 \times 10^{10}$ |
| $H_3O^+$ | 2·1–4·3 | $2·06 \times 10^{10}$ |
| $K^+$ | – | $< 5 \times 10^5$ |
| $N_2O$ | 7 | $8·7 \times 10^9$ |
| $O_2$ | 7 | $1·9 \times 10^{10}$ |
| $CH_3CO_2H$ | 5·4 | $1·8 \times 10^8$ |
| $CH_3COCH_3$ | 7 | $5·9 \times 10^9$ |
| $ClCH_2CO_2H$ | 1·0–1·5 | $6·9 \times 10^9$ |
| $CH_3CH_2OH$ | – | $< 4 \times 10^2$ |
| $C(NO_2)_4$ | 6 | $4·6 \times 10^{10}$ |
| Adenine | 6 | $3 \times 10^{10}$ |
| Purine | 7·2 | $1·7 \times 10^{10}$ |

The principal features of the reactions of the hydrated electron are:
1. The hydrated electron is a very powerful reductant. The alkali and alkaline-earth metal ions are generally unreactive though most other cations are reactive. The higher the positive charge on the cation, the greater is the reactivity. Inorganic cations, in general, are more reactive than inorganic anions. Rate coefficients are often very high and in some cases correspond to those for the diffusion-controlled reaction.
2. Aliphatic compounds are reactive only if they contain the groups C-Halogen, C=O, C=S, or S−S.
3. In its reaction with aromatic compounds the hydrated electron behaves as a typical nucleophilic reagent. Its reactivity thus depends on the $\pi$-electron density in the aromatic ring.

In very pure water the hydrated electron disappears via the reactions

$$e_{aq}^- + H_2O \rightarrow H\cdot + OH^- \tag{3.19}$$

$$e_{aq}^- + e_{aq}^- \rightarrow H_2 + 2OH^- \tag{3.20}$$

$$e_{aq}^- + H_3O^+ \rightarrow H\cdot + H_2O \tag{3.16}$$

the rate coefficients being 16, $5 \times 10^9$ and $2·06 \times 10^{10}$ $\mathrm{M^{-1}s^{-1}}$ respectively. At pH > 9, the contribution of reaction (3.16) is negligible. It is of interest to calculate the concentration of hydrated electrons that may be produced. The earliest experiments used pulses of about $10^{-6}$ s and with energies of about 2 rad per pulse.

$$\text{Since } 1 \text{ rad} = 6\cdot24 \times 10^{13} \text{ eV g}^{-1}$$
$$\text{Dose} \approx 1\cdot2 \times 10^{14} \text{ eV g}^{-1}$$

If $G(e_{aq}^-) \approx 3$ then the initial concentration of hydrated electrons is given by

$$[e_{aq}^-] = 1\cdot2 \times 10^{14} \times \frac{3}{10^2} \times \frac{10^3}{6 \times 10^{23}}$$
$$= 6 \times 10^{-9} \text{ M.}$$

The extinction coefficient is $1\cdot58 \times 10^4$ $\text{M}^{-1}\text{cm}^{-1}$ at 720 nm and, using a multiple absorption cell, significant absorption can be observed. At this low concentration at pH > 9, the hydrated electron will react via equation (3.19) with a typical first-order decay. The half life of the first-order reaction is given by

$$t_{1/2} = \frac{\log_e 2}{k}$$

where $k$ is the pseudo-first-order rate coefficient and is equal to $k_{3.19}[H_2O]$. The half life of the hydrated electron then is given by

$$t_{1/2} = \frac{\log_e 2}{16 \times 55} \simeq 8 \times 10^{-4} \text{ s.}$$

since $[H_2O] \simeq 55\text{M}$. Clearly this intermediate could be detected only if the pulse lifetime was significantly less than the electron lifetime and it is for this reason that the discovery of the hydrated electron awaited the development of accelerators capable of delivering microsecond pulses. More recently it has been possible to obtain pulses of higher energies and thus to obtain higher electron concentrations. It is interesting that measurements using pulses of 20 picoseconds duration have confirmed the absorption spectrum obtained at microsecond times indicating that, at least over this time region, the structure of the hydrated electron is constant. Reaction (3.20) probably occurs at the higher concentrations of hydrated electrons existing in the spurs.

The spectrum of the hydrated electron can be observed in continuous radiolysis if very powerful sources capable of delivering about 1000 rad s$^{-1}$ are used. Under these conditions the rate of production, $R$, of the hydrated electron is given by

$$R = 6\cdot24 \times 10^{13} \times 10^3 \times 10^3 \times \frac{3}{10^2} \times \frac{1}{6 \times 10^{23}}$$
$$\simeq 3 \times 10^{-6} \text{ M s}^{-1}$$

The stationary concentration is given by

$$\frac{d[e_{aq}^-]}{dt} = R - k[e_{aq}^-]_s[H_2O] = 0$$

i.e.

$$[e_{aq}^-]_s = \frac{3 \times 10^{-6}}{16 \times 55} \simeq 3 \times 10^{-9} \text{ M}$$

Thus its absorption has been measured using a multiple reflection cell in the radiolysis in a 15000 Ci $^{60}$Co source of $10^{-3}$ M aqueous sodium hydroxide saturated with hydrogen.

Although the existence of the hydrated electron has been well established, the fate of the counter ion, $H_2O^+$, is much less well established. Its lifetime in the gaseous phase is known to be extremely short because of the reaction

$$H_2O^+ + H_2O \rightarrow H_3O^+ + \cdot OH \qquad (3.11)$$

and it is supposed, though it is by no means certain, that its fate in liquid water will be the same. The production of ionic species in water radiolysis has also been established from measurements of the transient conductivity of pulse-irradiated water.

Solvated electrons have been observed in the pulse radiolysis of other polar liquids, e.g. alcohols. Absorption maxima are in the range 600–800 nm. The yields are lower than in water ($G \approx 1$ compared with $G \approx 3$ in water) although higher yields can be detected at high scavenger concentrations.

*Radiolysis of non-polar liquids*

Solvated electrons are much less likely to be observed in non-polar liquids for two reasons. The attractive force, $F$, between the electron and its parent ion is given by the Coulomb law

$$F = e^2/4\pi\varepsilon_r r^2$$

where $r$ is the interionic separation, $\varepsilon_r$ the relative permittivity (or dielectric constant) of the medium and $e$ is the electronic charge. In the non-polar liquid of lower relative permittivity, a thermalized electron is much more likely then to be drawn back by the Coulombic field of the parent positive ion. Even if the electron were to escape the parent ion, it is much less likely to be trapped since dipolar interaction with the solvent molecules will be negligible. Recombination between an electron and its parent ion will be complete within $10^{-10}$ s and is usually referred to as geminate recombination. Ions and electrons that survive geminate recombination and consequently escape into the bulk of the solvent are referred to as free ions and electrons. It is convenient to distinguish between the yield of geminate ions, $G_{gi}$, and the yield of free ions, $G_{fi}$. The energy $W$ required to create an ion pair is not greatly dependent on the substrate in the gaseous phase (see p. 23.) If it is assumed that $W$ is the same for liquids then the total ionic yield will be the same—$G_{fi} + G_{gi} \approx 3$, for both polar and non-polar liquids.

The yield of free ions has been determined from measurements of the conductivity of irradiated liquids. For a wide range of paraffin hydrocarbons of varying viscosity, values of $G_{fi} \approx 0.1$ were obtained. Clearly most electrons undergo geminate recombination. Similar values of $G_{fi}$ were obtained from microsecond pulse radiolysis of solutions containing aromatic solutes. When anthracene was used as a scavenger, a transient absorption peak at 730 nm was observed. This absorption is typical of the anthracene radical ion which had previously been prepared by the reaction of sodium with anthracene. Its extinction coefficient was known so that the yield of radical ions could be determined. Since the lifetime of the pulse was about $10^{-6}$ s only those ions existing after this time interval were measured. The spectrum of the positively charged anthracene ion radical is similar so that the total yield of positive and negative ions was measured. Results are shown in Figure 3.4.

Much higher ion yields can be obtained using other scavengers. $N_2O$ is known to be a specific electron scavenger in water radiolysis.

$$e_{aq}^- + N_2O \rightarrow N_2 + O^-$$  (3.21)

When $N_2O$ saturated solutions of cyclohexane were irradiated with $^{60}Co$ $\gamma$-rays, values of $G(N_2) = 4$ were obtained. This much higher yield is probably due to capture of geminate electrons by the $N_2O$.

As in the aqueous system, the reactions of positive ions in organic liquids have been much less characterized. It is well known that the polymerization of isobutene can proceed via a cationic mechanism. When isobutene is irradiated with 2 MeV electrons at 193 K, polymerization is observed. This probably results from the production of the $(CH_3)_3C^+$ ion from the reactions

$$i\text{-}C_4H_8 \xrightarrow{\ \sim\sim\ } C_4H_8^+ + e^-$$  (3.22)

$$C_4H_8^+ + i\text{-}C_4H_8 \rightarrow (CH_3)_3C^+ + CH_2\!=\!C\!-\!CH_2\cdot$$

$$\underset{CH_3}{\mid}$$  (3.23)

The ion has been identified in the pulse radiolysis of isobutene. E.s.r. studies indicate that in many organic liquids, the cation eliminates a proton to give a free radical.

$$C_6H_{14}^+ \rightarrow C_6H_{13}\cdot + H^+ \qquad (3.24)$$

The polymerization of olefins by radiation has been discussed by Jackson (1972).

**Radiolysis of solids**

Electrons have also been observed in the radiolysis of solid systems. Radiolysis of the alkali halides gives rise to absorption bands in the visible and ultraviolet regions. These are due to electrons which have been trapped in negative-ion vacancies in the lattice and are known as F-centres. The trapped electrons give characteristic e.s.r. spectra. Positive charges may also be trapped in cation vacancies and gives rise to V-centres.

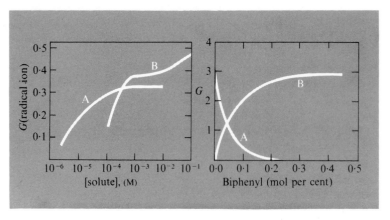

FIG. 3.4    Dependence of radical ion yields on solute concentration in irradiated liquid cyclohexane
A. anthracene as solute
B. benzophenone as solute
FIG. 3.5    Effect of biphenyl concentration on $G$(biphenyl anion) and $G$(trapped electron) for the radiolysis of solutions of biphenyl in ethanol at 77 K
A. $G$(trapped electron)
B. $G$(biphenyl anion)

A great deal of work has been carried out on the radiolysis of frozen liquids at about 77 K. At this low temperature many intermediates, which at room temperature are so short lived as to be detected only by pulse radiolysis, now have such long lifetimes that it is possible to produce them by continuous radiolysis in conventional sources and to characterize them by the usual spectrophotometric and e.s.r. techniques. The naphthalene and biphenyl anions have been identified in the radiolysis of glasses of tetrahydro-2-furan, ethanol and hydrocarbons containing naphthalene and biphenyl.

In selected systems it is possible to observe both the spectrum of the trapped electron and that of the anion formed by electron capture by an added solute. Some results for the radiolysis of biphenyl in an ethanol glass are shown in Figure 3.5. The reduction in $G$ (trapped electron) is accompanied by a corresponding increase in $G$ (anion).

There is much less evidence for the trapping of positive ions though these species have been observed in a limited number of systems.

## PROBLEMS

3.1. What concentration of hydrogen peroxide in air-saturated, neutral, aqueous solution ($[O_2] \simeq 2 \times 10^{-4}$M) would be required to react with 99 per cent of the hydrated electrons produced in the system?

3.2. Calculate the half life of the hydrated electron in a solution of initial concentration $10^{-8}$M at pH 3.

3.3 Calculate the initial concentration of hydrated electrons that would need to be produced so that the initial rates of disappearance by reactions (3.19) and (3.20) would be the same.

# 4. Excited states

### Production and decay of excited states

IN the early days of radiation chemistry excited states were convenient scapegoats by which radiation chemists sought to explain awkward observations! There was little direct evidence for their existence. However with the advent of pulse radiolysis it has been possible to provide more definite evidence for such species and to characterize some of them in detail. The knowledge thus gained is of importance not only in radiation chemistry but also in other fields, notably photosynthesis, vision, chemiluminescence and electrochemiluminescence where the participation of excited states is important.

In an irradiated system, excited states are produced either by direct excitation

$$M \quad \longrightarrow \quad M^* \tag{4.1}$$

or by ion neutralization.

$$M \quad \longrightarrow \quad M^+ + e^- \tag{4.2}$$

$$e + M^+ \quad \to \quad M^* \tag{4.3}$$

Excitation by high-energy electrons produces only singlet excited states whereas both singlet excited and triplet states are produced by slow electrons. Charge neutralization leads to the production of both singlet excited and triplet states†.

In a single-component system, the principal ways in which an excited state may dissipate its energy may be represented by the equations:

$$M_S \quad \to \quad M_G + h\nu \tag{4.4}$$

$$M_S \quad \to \quad M_G \tag{4.5}$$

$$M_S \quad \to \quad M_T \tag{4.6}$$

$$M_T \quad \to \quad M_G + h\nu \tag{4.7}$$

$$M^* \quad \to \quad products \tag{4.8}$$

$M_S$, $M_T$ and $M_G$ refer to the singlet excited, triplet and ground states respectively of M. $M^*$ is any excited state of M.

In reaction (4.4), the singlet excited state reverts to the ground state by the

---

† In the singlet excited state, an electron from an electron pair is excited to a higher energy orbital without any change in its direction of spin. In the triplet state, excitation of the electron is accompanied by inversion of its spin.

emission of one quantum of radiation. This process is known as fluoresence and occurs when the arrangement of potential-energy curves is as shown in Figure 4.1.

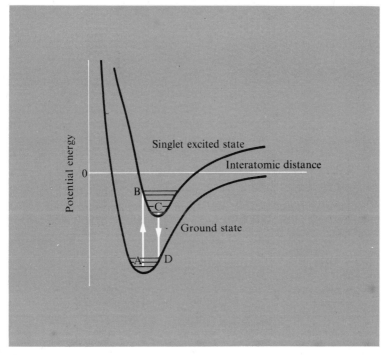

FIG. 4.1    Potential energy curves for a diatomic molecule. Excitation occurs from A to B and fluorescence from C to D

Fluorescence lifetimes are relatively short, about $10^{-8}$ s. The emission of fluorescence in an irradiated system is therefore indicative of the production of singlet excited states, because triplets decay more slowly (see below).

If the arrangement of potential-energy curves is as shown in Figure 4.2, then it is possible for the excited state to revert to the ground state without the emission of radiation, as in reaction (4.5). In this case the energy of the excited state appears as vibrational energy of the ground state. This process is known as internal conversion.

If the potential-energy curve for the singlet excited state crosses that of the triplet state then intersystem crossing to the triplet state may occur as in reaction (4.6). This is represented diagrammatically in Figure 4.3.

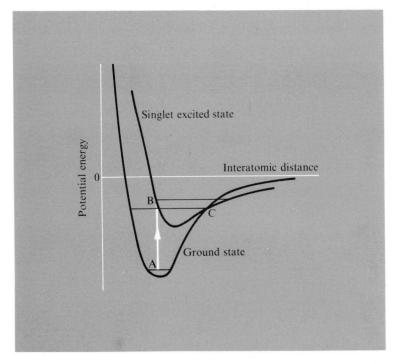

FIG. 4.2    Potential energy curves for a diatomic molecule. Excitation occurs from A to B and fluorescence for C to D

The triplet state decays to the ground state with the emission of radiation as in reaction (4.7). This process is known as phosphorescence. The relative arrangement of the potential-energy curves is similar to that for fluorescence of the singlet excited state. Triplet states however have much longer lifetimes, about $10^{-4}$ s, and they are therefore somewhat more easy to study than singlet excited states. Unlike singlet excited states, triplet states have longer lifetimes in more viscous media. The emission of phosphorescence in an irradiated system is therefore indicative of the production of triplet states.

Where a potential-energy curve corresponding to a repulsive state crosses that of the excited state, dissociation to give products can occur as in reaction (4.8). This arrangement is shown in Figure 4.4. The products may be either two free radicals or two stable molecules. Reaction (4.8) is the only reaction that leads to chemical change in the irradiated system.

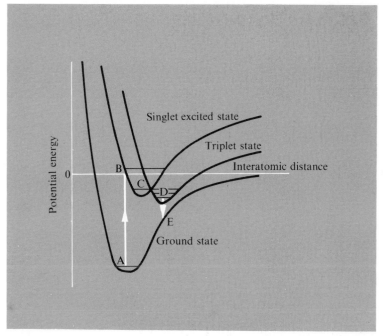

FIG. 4.3    Potential energy curves for a diatomic molecule. Excitation occurs from A to B and intersystem crossing at C. Phosphorescence is from D to E

In a multicomponent system energy transfer from one component to another can occur.

$$M^* + X \rightarrow M + X^* \qquad (4.9)$$

Processes of this type are important. In a dilute solution of a component X, most of the energy will be absorbed by the solvent, M. If energy transfer is efficient, the products observed will be those arising from the solute and not the solvent. Clearly for energy transfer to take place at all, the excitation energy of X must be less than that of M. The transfer of energy will be most efficient when the excitation energies of X and M are similar.

### Evidence for the existence of excited states

*Triplet state*

The spectrum obtained on the pulse radiolysis of a dilute solution of anthracene in acetone is shown in Figure 4.5. This spectrum is identical to that observed for the triplet state of anthracene produced in flash photolysis

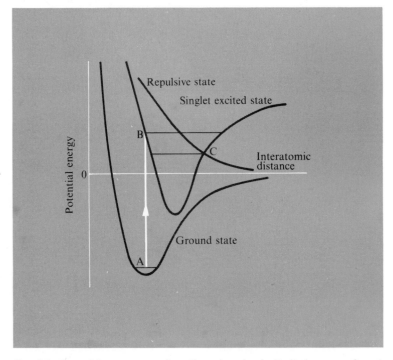

FIG. 4.4   Potential energy curves for a diatomic molecule. Excitation occurs from A to B and dissociation at C

studies and clearly indicates the existence of this species in the radiolysis. In the dilute solution all the energy of the radiation is absorbed by the acetone and the anthracene triplet must arise via an energy transfer process. The rate of formation of the anthracene triplet indicated that it was produced from an acetone precursor which itself had a lifetime of more than 5 $\mu$s for a first-order decay in pure acetone. This long lifetime suggests that the acetone precursor is itself a triplet rather than a singlet excited state and the reaction is

$$(CH_3COCH_3)_T + C_{14}H_{10} \rightarrow CH_3COCH_3 + (C_{14}H_{10})_T \quad (4.10)$$

The rate coefficient for the above reaction was found to be $6 \cdot 2 \times 10^9$ M$^{-1}$s$^{-1}$ at 296 K. Scavenger studies indicated that $G$ (acetone$_T$) $\approx 1 \cdot 1$.

The triplet state of benzene ($^3B_{1u}$) has been identified in the pulse radiolysis of liquid benzene. The yield of triplet states has been determined in two ways. *cis*–Butene is isomerized to *trans*–butene by reaction with the benzene triplet.

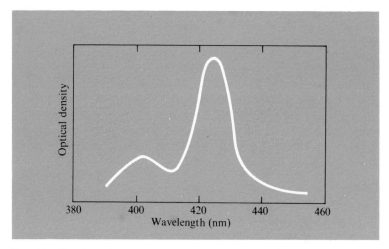

Fig. 4.5 Transient absorption produced in the pulse radiolysis of solutions of anthracene in acetone

The yield of *cis–trans* isomerization in the continuous radiolysis of solutions of *cis*–butene in benzene is thus a measure of the yield of benzene triplet. In pulse radiolysis the production of triplet states of added solutes—e.g. naphthalene, biphenyl, and anthracene—by energy transfer from the benzene triplet can be observed and, since the extinction coefficient of the solute triplet is known from flash-photolysis studies, the yield of benzene triplet can be deduced. In these systems allowance must be made for the fact that some solute triplet is formed by intersystem crossing from the solute singlet excited state produced by transfer from a benzene singlet excited state. The yields observed by both methods are in good agreement and give $G\,(\text{benzene}_T) = 4\cdot2$.

The half life of the benzene triplet in liquid benzene is $2\cdot3$ ns. A transitory species with a half life of formation of about 3 ns and a decay half life of about 112 ns has also been observed. From the effect of scavengers, it was concluded that the species was a biradical of benzene produced by interaction of triplet benzene with its ground state.

*Singlet excited state*

There is much less direct evidence for the existence of the singlet excited state in radiolysis. This is to be attributed to the experimental difficulty of detecting a species of such short half life and is not to imply that the singlet excited state is not produced.

Using nanosecond pulse radiolysis both singlet excited and triplet states have been observed in the radiolysis of anthracene in cyclohexane. The spectra of singlet excited and triplet anthracene are similar. The yields of these excited states are dramatically reduced but not entirely eliminated by the addition of scavengers able to react with positive ions, e.g., alcohols and amines, and by electron scavengers, e.g. $N_2O$ and $CH_3I$, indicating that although the major part of these yields arises from charge neutralization, a small but significant part arises by direct excitation. Similar results were obtained in the radiolysis of benzene solutions. Whereas almost all of the yield of triplet benzene was removed by ion scavengers, only about 60 per cent of the yield of singlet excited benzene $(G(\text{singlet}) = 1\cdot2)$ was removed.

With the advent of picosecond pulse radiolysis, it is to be hoped that much valuable information regarding the singlet excited state will be forthcoming.

### Chemical reactions of excited states

Indirect chemical evidence for the participation of excited states in radiolysis was obtained long before the direct confirmation by pulse radiolysis of their existence. A classic example is the radiolysis of cyclohexane–benzene solutions. The hydrogen yields for the 1 MeV electron radiolysis of liquid cyclohexane and benzene are 5·8 and 0·036 respectively. In a mixture of the two components it might be expected that, since energy will be absorbed by each component in direct proportion to its electron fraction, the observed hydrogen yield would be given by

$$G(H_2) = 5\cdot8e_C + 0\cdot036e_B \qquad (4.11)$$

where $e_C$ and $e_B$ are the electron fractions of cyclohexane and benzene respectively. In fact $G(H_2)$ is always lower than the expected value, as may be seen from Figure 4.6.

In Figure 4.6 the dotted line is the theoretical yield according to equation (4.11). Substitution of benzene by hexadeuterobenzene showed clearly that the decrease in $G(H_2)$ was not due to the reaction

$$H\cdot + C_6H_6 \rightarrow C_6H_7\cdot \qquad (4.12)$$

The simplest explanation is that an excited cyclohexane molecule transfers its energy to benzene and is therefore prevented from decomposing. Alternative explanations have been advanced. The concept of 'sponge type' protection, as displayed by benzene, is important and has stimulated work in radiobiology. Thus it is possible that the harmful effects of radiation on a living cell might be lessened by the addition of a suitable protective agent, which could accept excitation energy from the cell components but itself be relatively stable.

Phosphorescence is observed in the photolysis of liquid cyclopentanone,

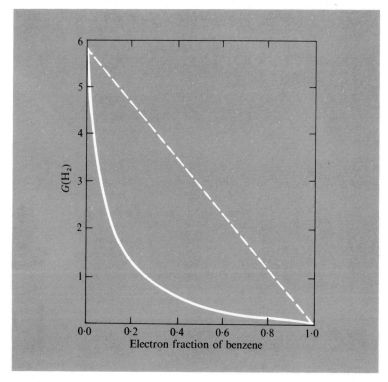

FIG. 4.6   Dependence of $G(H_2)$ on the electron fraction of benzene for the radiolysis of cyclohexane-benzene solutions.
———————— observed $G(H_2)$
– – – – – –   predicted $G(H_2)$ if no energy transfer

indicating the formation of the triplet state. 4-Pentenal is an important product and appears to be produced from the triplet state. The same product is observed in the radiolysis and although its yield is unaffected by conventional radical scavengers, e.g. diphenylpicrylhydrazyl and iron(III) chloride, it is decreased by oxygen or 1,3-pentadiene, which are known to react with triplet states. The effect of 1,3-pentadiene is consistent with a reaction mechanism:

$$\text{cyclopentanone} \; + \; h\nu \; \longrightarrow \; \left( \text{cyclopentanone} \right)_S \qquad (4.13)$$

$$\left(\begin{array}{c} \text{(cyclopentanone)} \\ \text{O} \end{array}\right)_S \rightarrow \left(\begin{array}{c} \text{(cyclopentanone)} \\ \text{O} \end{array}\right)_T \qquad (4.14)$$

$$\left(\begin{array}{c} \text{(cyclopentanone)} \\ \text{O} \end{array}\right)_T \longrightarrow CH_2 = CHCH_2CH_2CHO \qquad (4.15)$$

$$\left(\begin{array}{c} \text{(cyclopentanone)} \\ \text{O} \end{array}\right)_T + C_5H_8 \longrightarrow \begin{array}{c} \text{(cyclopentanone)} \\ \text{O} \end{array} + (C_5H_8)_T$$

$$(4.16)$$

$k_{4.16}/k_{4.15}$ was observed to be 21 $M^{-1}$ both in the photolysis and in the radiolysis.

Although excited states have frequently been postulated in the radiolysis of water there is, as yet, no direct evidence for the existence of an excited water molecule.

### Excited states in the radiolysis of gases

In Chapter 3 it was seen that the contributions of ionic and excited states to the reaction could be separated by applying an electric field to the sample being irradiated. This important technique has been further extended to give very valuable results. If the applied field is relatively low then electrons will be accelerated but the lifetimes of ions will still be sufficiently long to permit them to undergo ion–molecule reactions before being collected at the electrodes. Under these conditions it might be expected that the yields of products from excited states should increase in so far as collisions of the accelerated electrons with the substrate will increase the yield of excited states. The yield of products arising from ion–molecule reactions would be expected to remain the same.

From the study of the photolysis of ethane, it was shown that excited ethane decomposes by the following pathways:

$$C_2H_6^* \rightarrow H_2 + CH_3CH: \qquad (4.17)$$

$$C_2H_6^* \rightarrow C_2H_4 + H\cdot + H\cdot \qquad (4.18)$$

$$C_2H_6^* \rightarrow CH_4 + CH_2: \qquad (4.19)$$

The relative importance of each of these modes is determined by the energy of the u.v. quantum. If a mixture of $C_2H_6$ and $C_2D_6$ is radiolysed, then the products arising from excited molecule reactions will be isotopically simple. Isotopically mixed compounds of the type $CD_3H$ can be obtained only as a result of ion–molecule reactions. Thus in the $\gamma$-radiolysis of $C_2H_6$–$C_2D_6$–NO mixtures (nitrogen monoxide was added to scavenge all free radicals) the yields of $C_2D_4$ and $CD_4$ increased nearly fourfold over the range 0–1300 V whereas the yields of $C_2D_3H$ and $CD_3H$ remained constant. This distribution of products is in excellent agreement with the above predictions. Comparison with the photolysis results indicated that the average energy imparted to an ethane molecule on electron impact was > 10 eV.

In gaseous hydrocarbons the yields of excited molecules are usually less than the yields of ion pairs. The ratios $G$(excited molecules)$/G$(ion pairs) for n-propane, ethane and ethylene are 0·3, 0·5 and 1·0 respectively. These observations are consistent with results for the radiolysis of non-polar liquids where it was seen that the yield of excited states produced by direct excitation is usually less than the yield produced by charge neutralization.

PROBLEMS

4.1. If the only reaction leading to the decay of the benzene triplet is

$$(C_6H_6)_T \quad \rightarrow \quad \text{products}$$

calculate the first-order rate coefficient for the above reaction given that the half life of the triplet state is 2·3 ns. If $G$(benzene$_T$) is 4·2, calculate the yield of anthracene triplet in a benzene solution containing $5 \times 10^{-3}$M anthracene given that the rate coefficient for the reaction

$$(C_6H_6)_T + C_{14}H_{10} \quad \rightarrow \quad C_6H_6 + (C_{14}H_{10})_T$$

is $3·9 \times 10^{10}$ M$^{-1}$s$^{-1}$.

4.2. Derive an expression for the relationship between $G$(4-pentenal), $G$(cyclopentanone$_T$) and 1,3-pentadiene concentration for the radiolysis of cyclopentanone. Hence calculate the fraction of cyclopentanone triplet converted to 4-pentenal at [1,3-pentadiene] = 0.1M.

4.3. Calculate the minimum energy for an electron (in eV) required to excite a gaseous ethane molecule such that it might decompose via reaction (4.19). The heats of formation, $\Delta H_f^{\ominus}$, of $CH_2$, $CH_4$ and $C_2H_6$ at 298 K are 398, −74·9 and −84·5 kJ mol$^{-1}$ respectively.

# 5. Free radicals

## Production and detection of free radicals in irradiated systems

*Production of free radicals*

IN Chapter 4 it was seen that the decomposition of an excited molecule can lead to the production of free radicals. Free radicals then are seldom, if ever, primary products of the radiolysis but arise from subsequent decomposition of the ions and excited states produced initially. Typical reactions are:

$$C_6H_{14}^* \rightarrow C_6H_{13}\cdot + H\cdot \tag{5.1}$$

$$C_2H_6^+ \rightarrow C_2H_5\cdot + H^+ \tag{5.2}$$

$$e^- + CH_3Cl \rightarrow CH_3\cdot + Cl^- \tag{5.3}$$

The essential property of a free radical is its possession of one or more unpaired electrons. The measurement of their electron spin resonance is therefore the most conclusive demonstration of the existence of free radicals in a system.[†] Some free radicals may however be detected by characteristic absorption spectra. Free radicals are usually extremely reactive species. At low concentrations, they tend to react with other molecules, e.g.

$$R\cdot + R^1H \rightarrow RH + R^1\cdot \tag{5.4}$$

to produce a new and more stable radical in the system. At higher concentrations, they may react with each other to give a stable molecule.

$$R\cdot + R\cdot \rightarrow R_2 \tag{5.5}$$

The characterization of free radicals in irradiated systems has been by two different techniques. Historically, the first and simpler technique was to produce the radicals in the irradiated solid at low temperatures, about 77 K. Under these conditions the radical was much less reactive since it was essentially trapped in the solid. High concentrations could be built up in the solid by prolonged irradiation even with relatively low-intensity facilities. The radicals could then be characterized at leisure by e.s.r. and other spectroscopic techniques. More recently free radicals have been detected in liquids and gases either by using the technique of pulse radiolysis to achieve an initially high concentration of radicals, or by using very high-intensity irradiation facilities.

---

[†] For a good introduction to the basic principles of electron spin resonance, the reader should consult McLauchlan (1972).

*Electron spin resonance spectra of radicals*

The first e.s.r. observations on radiation produced radicals were published in 1954. Irradiation of frozen aqueous solutions of $H_2SO_4$, $HClO_4$, and $H_3PO_4$ at 77 K produced a doublet in the e.s.r. spectrum, characteristic of the hydrogen atom. That this was indeed due to the hydrogen atom was confirmed by the observation that substitution of $H_2O$ by $D_2O$ led to the formation of three new lines (the nuclear spin of the deuterium atom is 1). When the solution was warmed the e.s.r. signal disappeared. Study of the rate of disappearance at a given temperature led to the conclusion that the hydrogen atoms disappeared by a second-order process without developing electrical conductivity, i.e.

$$H\cdot + H\cdot \rightarrow H_2 \tag{5.6}$$

It is interesting that no e.s.r. spectrum attributable to the hydrogen atom was observed in pure ice or in frozen HCl solutions at 77 K although the spectrum was observed in ice at 4 K. It appears that in the relatively open ice structure the hydrogen atom still has sufficient mobility at 77 K to undergo recombination. There must be some interaction with the anions $SO_4^{2-}$, $ClO_4^-$ and $PO_4^{3-}$ (but not $Cl^-$) which leads to stabilization of the hydrogen atom. The hydroxyl radical is somewhat less reactive and can be detected in ice at 77 K.

A similar doublet is observed in the radiolysis of many organic compounds at 77 K or below. Thus in Figure 5.1, the e.s.r. spectrum of methane at 4 K shows the characteristic hydrogen-atom doublet and the methyl radical quartet.

The e.s.r. spectra of radicals trapped in poly-crystalline solids are not very well resolved and although the spectra obtained unequivocally demonstrate the existence of free radicals, it is difficult to analyse the spectra for structural details. The irradiation of single crystals was found to give relatively simple but orientation-dependent spectra. From measurements of the magnetic anisotropy of these spectra, it was concluded that the radical retained the molecular orientation of the host crystal. This type of study has been described by McLauchlan (1972).

Well resolved spectra are obtained from radicals in the liquid phase. E.s.r. spectrometers are capable of measuring radical concentrations down to about $10^{-8}$ M and this is about the stationary concentration that can be achieved using high dose rates. Thus it has been possible to measure e.s.r. spectra *in situ* by directing a high intensity electron beam from a Van de Graaff accelerator through an axial hole in the magnet of the spectrometer to the sample in the cavity of the magnet. The spectrum of the ethyl radical obtained from the radiolysis of liquid ethane is shown in Figure 5.2. The

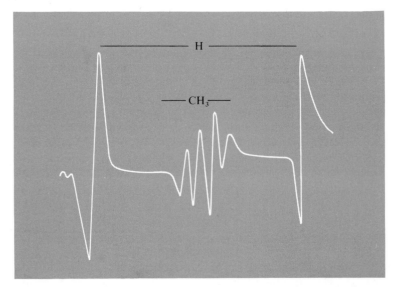

FIG. 5.1    Electron spin resonance spectrum of γ-irradiated solid methane at 4 K

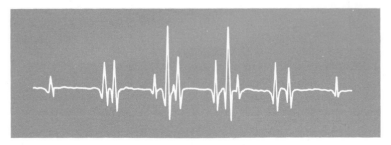

FIG. 5.2    Electron spin resonance spectrum of electron irradiated liquid ethane at 93 K

12-line spectrum is entirely consistent with that expected for such a radical. The technique is ideally suited to the study of non-polar liquids; it is much less satisfactory for polar substances, since the high dielectric loss of such materials severely restricts the size of sample that may be studied. A wide range of hydrocarbons has been investigated in this way and the results are important, not only for demonstrating the existence of radical species in these systems, but also for the detailed information that can be derived from the spectra on the electron distribution in these radicals. Collections of spectra have been published.

Although the ethyl radical is the major radical in the radiolysis of liquid ethane, lines attributable to the methyl and vinyl radicals were observed in the e.s.r. spectrum. The vinyl concentration was about 3 percent of the ethyl radical concentration and was unaffected by the addition of deuterio-ethylene which would be expected to scavenge any hydrogen atoms produced in the system and produce ethyl radicals.

$$C_2D_4 + H\cdot \rightarrow C_2D_4H\cdot \tag{5.7}$$

It was concluded that the vinyl radical does not therefore arise from decomposition of the ethyl radical

$$CH_3CH_2\cdot \nrightarrow CH_2{=}CH\cdot + H_2 \tag{5.8}$$

but is a primary product of the radiolysis.

$$CH_3CH_3 \rightsquigarrow CH_2{=}CH\cdot + H_2 + H\cdot \tag{5.9}$$

The stationary state concentration of ethyl radicals was shown to be proportional to the square root of the dose rate. This is consistent with a second-order recombination reaction for the disappearance of the radicals.

$$C_2H_6 \rightsquigarrow C_2H_5\cdot + H\cdot \tag{5.10}$$

$$C_2H_5\cdot + C_2H_5\cdot \rightarrow C_4H_{10} \tag{5.11}$$

By using the stable free radical galvinoxyl as a reference it was possible to evaluate the absolute radical concentration from the intensity of the spectrum and hence to calculate the rate coefficient for recombination. Thus $k_{5.11}$ was shown to be $3 \times 10^8$ $M^{-1}s^{-1}$ at 98 K and the activation energy of the reaction was 3·3 kJ mol$^{-1}$. This low value of the activation energy is consistent with a diffusion-controlled recombination reaction.

In higher hydrocarbons, the major radicals arise from loss of a hydrogen atom from the parent molecule although radicals arising from C-C bond rupture are observed in branched-chain hydrocarbons. In oxygenated solution, the hyperfine structure of the spectrum is lost and only a single broad line is observed. This is attributed to the alkyl peroxy radical.

$$R\cdot + O_2 \rightarrow RO_2\cdot \tag{5.12}$$

*Optical absorption spectra of radicals*

Although spectra of trapped radicals have been extensively measured, the more interesting results from the point of view of the radiation chemist have been obtained from the pulse radiolysis technique. The first application of this technique was by R. L. McCarthy and A. MacLachlan in 1960 and resulted in the identification of the benzyl radical. Its spectrum is shown in Figure 5.3. Identification of the radical was based on the absorption maxima

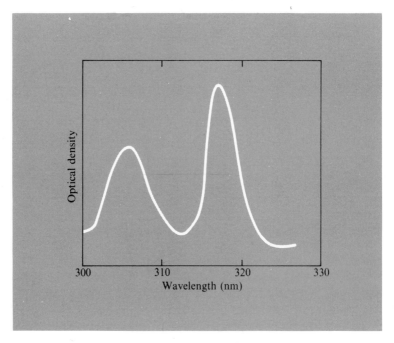

FIG. 5.3   Absorption spectrum of the benzyl radical

at 306 and 318 nm, the same as that for the benzyl radical which was known to be produced in the flash photolysis of benzyl chloride solutions. The spectrum was observed in the radiolysis of solutions of benzyl chloride, alcohol or formate in a solvent of 33 per cent ethanol and 67 per cent glycerol. No spectrum was observed, however, using cyclohexane as solvent. The apparatus used did not permit measurement of spectra before 25 $\mu$s after the pulse and it appears that in the less viscous cyclohexane, the radicals recombine before they can be detected.

In ethanol–glycerol solutions it was shown that the benzyl radicals decayed by two competing reactions

$$C_6H_5CH_2\cdot \ + \ C_6H_5CH_2\cdot \ \rightarrow \ (C_6H_5CH_2)_2 \qquad (5.13)$$

$$C_6H_5CH_2\cdot \ + \ CH_3\cdot CHOH \ \rightarrow \ C_6H_5CH_2CH(CH_3)\,OH \quad (5.14)$$

Although the absolute extinction coefficient of the benzyl radical was not known at the time it was possible by product analysis to arrive at an approximate value for the yield of benzyl radicals and hence to deduce that $\varepsilon_{318} \approx 1 \cdot 1 \times 10^3$ $M^{-1}cm^{-1}$. The rate coefficients for reactions 5·13 and 5·14 were $4 \times 10^7$ and $2 \times 10^8$ $M^{-1}s^{-1}$ respectively. These are the expected values for the diffusion-controlled bimolecular recombination.

In the pulse radiolysis of cyclohexane–oxygen solutions a transient absorption with maxima at 276 and 295 nm was observed, although no absorption was observed in this region in deoxygenated cyclohexane. The major oxidation products were identified as cyclohexanol, cyclohexanone and cyclohexyl hydroperoxide. It seems likely that the transient absorption is due to the cyclohexylperoxy radical and that the reaction mechanism is

$$C_6H_{12} \rightsquigarrow C_6H_{11}\cdot + H\cdot \tag{5.15}$$

$$H\cdot + O_2 \rightarrow HO_2\cdot \tag{5.16}$$

$$C_6H_{11}\cdot + O_2 \rightarrow C_6H_{11}O_2\cdot \tag{5.17}$$

$$2C_6H_{11}O_2\cdot \rightarrow C_6H_{11}OH + C_6H_{10}O + O_2 \tag{5.18}$$

$$C_6H_{11}O_2\cdot + HO_2\cdot \rightarrow C_6H_{11}O_2H + O_2 \tag{5.19}$$

The rate coefficients for reactions (5.18) and (5.19) were $1 \cdot 6 \times 10^6$ and $3 \times 10^6$ $M^{-1}s^{-1}$ respectively. These rate coefficients are approximately $10^{-4}$ times the diffusion-controlled rate coefficients but were independent of temperature, indicating that these low values arise because of steric hindrance to the recombination. It is interesting that similar values for the rate coefficients were obtained from photo-oxidation studies of cyclohexene.

## Free radicals in the radiolysis of water

### The hydroxyl radical

In the pulse radiolysis of water, in addition to the intense absorption in the range 400–1000 nm which has been attributed to the hydrated electron, a much less intense absorption is observed in the range 200–400 nm. The allocation of this spectrum to the hydroxyl radical is based on the following evidence:

1. It is reduced by scavengers known to react with hydroxyl radicals. The efficiencies of different scavengers in suppressing the spectrum is proportional to their rates of reaction with hydroxyl radicals as determined in other systems.
2. The intensity of the spectrum is increased when hydrated electrons are

50     FREE RADICALS

converted to hydroxyl radicals via the reactions

$$e_{aq}^- + N_2O \rightarrow N_2 + \cdot O^-$$ (5.20)

$$\cdot O^- + H_2O \rightarrow OH^- + \cdot OH$$ (5.21)

The absorption spectrum is shown in Figure 5.4.

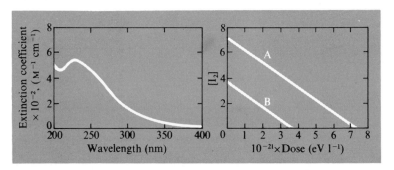

FIG. 5.4   Absorption spectrum of the hydroxyl radical in liquid water
FIG. 5.5   Disappearance of iodine in the $\gamma$-radiolysis of n-hexane–iodine solutions
A. initial $[I_2] = 9 \times 10^{-4}$ M
B. initial $[I_2] = 4.5 \times 10^{-4}$ M

Because the absorption is much less intense ($\varepsilon$ at 260 nm is about $4 \times 10^2$ $M^{-1}cm^{-1}$) and is in a much less convenient part of the spectrum for routine measurement, the system does not lend itself as readily as does the hydrated-electron system to the determination of rate coefficients for scavengers. There is much less kinetic data available for the hydroxyl radical. It should be possible to add to the system a solute that reacts with the hydroxyl radical to produce an intensely absorbing species in a more convenient part of the spectrum. The thiocyanate ion has been used.

$$CNS^- + \cdot OH \rightarrow CNS\cdot + OH^-$$ (5.22)

The product has a convenient absorption at 500 nm and $k_{5.22}$ has been measured. Addition of hydroxyl radical scavengers will suppress this absorption because of the competing reaction

$$\cdot OH + RH \rightarrow H_2O + R\cdot$$ (5.23)

and it is thus possible, in principle, to measure the relative rate ratio $k_{5.23}/k_{5.22}$ and hence to obtain $k_{5.23}$. However interpretation of the results is complicated by the reaction

$$CNS\cdot + CNS^- \rightarrow (CNS)_2^-\cdot$$ (5.24)

and the system is not entirely suitable for the determination of rate coefficients.

In aqueous benzene solution, the hydroxyl-radical–benzene adduct formed in the reaction

$$\cdot OH + C_6H_6 \rightarrow C_6H_6OH\cdot \qquad (5.25)$$

has an intense absorption and can be used for the determination of the rate coefficients for selected solutes.

Reactions of the hydroxyl radical are of three kinds:

1) hydrogen-atom abstraction, e.g.

$$CH_3OH + \cdot OH \rightarrow \cdot CH_2OH + H_2O \qquad (5.26)$$

2) electron-transfer reactions, e.g.

$$Fe^{2+} + \cdot OH \rightarrow Fe^{3+} + OH^- \qquad (5.27)$$

3) addition reactions, e.g.

$$CH_2 = CH_2 + \cdot OH \rightarrow HOCH_2CH_2\cdot \qquad (5.28)$$

In many cases rate data have had to be obtained by conventional continuous radiolysis using product analysis and competition kinetics (see p. 63). This is much more time consuming than pulse radiolysis and consequently much less popular! Some selected data are given in Table 5.1.

TABLE 5.1

*Rate coefficients for reactions of the hydroxyl radical at about 300 K*

| Reactant | pH | Rate coefficient ($M^{-1}s^{-1}$) |
|---|---|---|
| $e_{aq}^-$ | 11 | $3 \times 10^{10}$ |
| H· | – | $3.2 \times 10^{10}$ |
| $OH^-$ | – | $3.6 \times 10^8$ |
| $Br^-$ | 2 | $5 \times 10^9$ |
| $I^-$ | 2; 7 | $3.4 \times 10^{10}$ |
| $Fe(CN)_6^{4-}$ | 3–10 | $1.1 \times 10^{10}$ |
| $H_2$ | 7 | $6 \times 10^7$ |
| $CH_3OH$ | 7 | $9.5 \times 10^8$ |
| $C_2H_5OH$ | 7 | $1.6 \times 10^9$ |
| $CH_3CH(OH)CH_3$ | 7 | $1.5 \times 10^9$ |
| $C_6H_6$ | – | $4.3 \times 10^9$ |

That the oxidizing species in neutral aqueous solutions is the hydroxyl radical is confirmed by the observation that there is no kinetic salt effect on its reactions. Thus in oxygenated aqueous ethanol, hydrogen peroxide arises via the reactions

$$H_2O \quad \longrightarrow \quad H\cdot, OH\cdot, e_{aq}^-, H_2, H_2O_2 \qquad (5.29)$$

$$H\cdot + O_2 \quad \rightarrow \quad HO_2\cdot \qquad (5.30)$$

$$C_2H_5OH + \cdot OH \quad \rightarrow \quad CH_3\cdot CHOH + H_2O \qquad (5.31)$$

$$CH_3\cdot CHOH + O_2 \quad \rightarrow \quad CH_3CHO + HO_2\cdot \qquad (5.32)$$

$$2HO_2\cdot \quad \rightarrow \quad H_2O_2 + O_2 \qquad (5.33)$$

Addition of bromide ion decreases $G(H_2O_2)$ because of the reaction

$$Br^- + \cdot OH \quad \rightarrow \quad Br\cdot + OH^- \qquad (5.34)$$

The bromine atom does not abstract hydrogen from ethanol. Addition of inert electrolytes however does not affect the relative rate of reactions (5.31) and (5.34) indicating that the oxidizing radical must be uncharged. However at pH 13, the kinetic salt effect is consistent with a species of unit negative charge. This arises from the ionization

$$\cdot OH \quad \rightleftharpoons \quad O^-\cdot + H^+ \qquad (5.35)$$

and $pK_{OH}$ has been measured as 11·8 at 296 K.

*The hydrogen atom*

Spectral evidence for the existence of atomic hydrogen in water is much less well established. Pulse radiolysis of deaerated $10^{-3}$ M $HClO_4$ and 0·027 M $H_2$ shows the existence of a species absorbing in the region 200–240 nm. At 200 nm, $\varepsilon = 900$ M$^{-1}$cm$^{-1}$ but decreases to zero at 240 nm. Under the conditions of the radiolysis, both hydrated electrons and hydroxyl radicals would be rapidly converted to hydrogen atoms.

$$e_{aq}^- + H_3O^+ \quad \rightarrow \quad H\cdot + H_2O \qquad (5.36)$$

$$\cdot OH + H_2 \quad \rightarrow \quad H\cdot + H_2O \qquad (5.37)$$

Although it is difficult to suggest an alternative species if the absorption is not due to the hydrogen atom, it is somewhat surprising that the spectrum is so strongly shifted towards the red. The gaseous hydrogen atom in its ground state does not absorb light of wavelength longer than 121 nm.

There is however considerable kinetic evidence in support of the existence of the hydrogen atom. Thus the existence of two distinct reducing species in the radiolysis of water is well established. The species have kinetically different behaviour with a range of solutes, e.g. $ClCH_2CO_2H$, Fe(III), $H_2O_2$. These two species are interconvertible and the species in neutral solution has been shown unequivocally to be the hydrated electron. The other species, predominant at low pH, is most likely therefore to be the hydrogen atom.

$$H_3O^+ + e_{aq}^- \rightleftharpoons H\cdot + H_2O \tag{5.38}$$

At pH 2, the kinetic salt effect indicates that the reducing agent is uncharged.

Although the conversion of the hydrated electron to the hydrogen atom, as in reaction (5.36), is rapid ($k_{5.36}$ is $2\cdot06 \times 10^{10}$ $M^{-1}s^{-1}$), the conversion of the hydrogen atom to the hydrated electron

$$H\cdot + OH^- \rightarrow e_{aq}^- + H_2O \tag{5.39}$$

is relatively slow ($k_{5.39}$ is $2 \times 10^7$ $M^{-1}s^{-1}$). That this reaction undoubtedly occurs was demonstrated by passing hydrogen atoms, produced by a microwave discharge in hydrogen, into aqueous alkaline solution when they showed the characteristic behaviour of hydrated electrons. Further evidence for the conversion was obtained from pulse radiolysis studies of water saturated with hydrogen at high pressure, about 100 atm. The concentration of hydrogen was approximately $0\cdot1$ M and under these conditions all OH radicals react via (5.37). From pH 7–11 the initial optical absorption is constant and corresponds to $G_e$.[†] However at pH 12–13 there occurs a marked increase in the absorption. This is to be attributed to the conversion of hydrogen atoms, which are produced both as a primary product and by reaction (5.37), to hydrated electrons as in reaction (5.39). The ratio of optical densities should be given by

$$\frac{\text{O.D. at pH 7}}{\text{O.D. at pH 13}} = \frac{G_e}{G_e + G_H + G_{OH}}$$

The experimental value of the ratio is 0.46 and is in good agreement with that calculated from primary yields (see p. 59).

It is interesting to calculate the acid dissociation constant of the hydrogen atom. Thus the equilibrium constant for the equilibrium

$$H\cdot + OH^- \rightleftharpoons e_{aq}^- + H_2O \tag{5.40}$$

is given by

$$K = \frac{[e_{aq}^-][H_2O]}{[H\cdot][OH^-]} = \frac{k_{5.39}}{k_{3.19}} = \frac{2 \times 10^7}{16} = 1\cdot25 \times 10^6$$

---

[†] It is customary to denote yields of primary products in a radiolysis by a suffix, e.g. $G_e$ but yields of secondary products by brackets, e.g. $G(Fe(III))$.

The acid dissociation constant is defined by

$$K_a = \frac{[H^+][e_{aq}^-]}{[H\cdot]}$$

$$\text{i.e. } K_a = \frac{K[H^+][OH^-]}{[H_2O]}$$

But for water at 298 K

$$[H^+][OH^-] \approx 10^{-14} M^2$$
$$\text{and } [H_2O] \approx 55 M$$
$$\therefore \quad K_a = \frac{2 \times 10^7}{16} \times \frac{10^{-14}}{55} M$$
$$= 2\cdot3 \times 10^{-10} M$$

This corresponds to $pK_a = 9\cdot6$ and the hydrogen atom thus behaves as a very weak acid.

The standard free-energy change for reaction (5.40) is given by

$$\Delta G^{\ominus} = -RT \ln K$$
$$= -34\cdot8 \text{ kJ mol}^{-1}$$

It is thus possible to calculate the standard free-energy change for the reaction

$$\tfrac{1}{2}H_2 \rightarrow H_3O^+ + e_{aq}^- \qquad (5.41)$$

by considering the following cycle

|  |  | $\Delta G^{\ominus}$ (kJ mol$^{-1}$) |  |
|---|---|---|---|
| $H_{aq}^{\cdot} + OH^-$ | $\rightarrow e_{aq}^- + H_2O$ | $-34\cdot8$ | (5.39) |
| $H_2O$ | $\rightarrow H_3O^+ + OH^-$ | $+79\cdot9$ | (5.42) |
| $\tfrac{1}{2}H_2$ | $\rightarrow H_g^{\cdot}$ | $+203\cdot3$ | (5.43) |
| $H_g^{\cdot}$ | $\rightarrow H_{aq}^{\cdot}$ | $+19\cdot3$ | (5.44) |
| Net $\tfrac{1}{2}H_2$ | $\rightarrow H_3O^+ + e_{aq}^-$ | $+267\cdot7$ | (5.41) |

This free-energy change of $+267\cdot7$ kJ mol$^{-1}$ corresponds to a standard electrode potential of $-2\cdot77$ V since $\Delta G^{\ominus} = -FE^{\ominus}$. In electrochemistry, electrode potentials have been arbitrarily measured against a value of zero for the standard hydrogen electrode, i.e. relative to a value for $\Delta G^{\ominus}$ for reaction (5.41) of 0. It is now possible to determine absolute values of these potentials by the addition of $+2\cdot77$ V.

Reactions of the hydrogen atom are of three kinds:

1) hydrogen-atom abstraction, e.g.

$$CH_3CH_2OH + H\cdot \rightarrow CH_3\cdot CHOH + H_2 \qquad (5.45)$$

2) electron-transfer reactions, usually as a reducing agent, e.g.

$$Cu^{2+} + H\cdot \rightarrow Cu^+ + H^+ \qquad (5.46)$$

but in special cases as an oxidizing agent e.g.

$$Fe^{2+} + H\cdot + H^+ \rightarrow Fe^{3+} + H_2 \qquad (5.47)$$

3) addition reactions, e.g.

$$C_6H_6 + H\cdot \rightarrow C_6H_7\cdot \qquad (5.48)$$

As with the hydroxyl radical, much of our knowledge of hydrogen-atom reactions has come from competition studies using continuous radiolysis and there is a similar scarcity of data. Some typical results are shown in Table 5.2.

TABLE 5.2

*Rate coefficients for reactions of the hydrogen atom at about 300 K*

| Reactant | pH | Rate coefficient $(M^{-1}s^{-1})$ |
|---|---|---|
| $e^-_{aq}$ | 10·5 | $2\cdot5 \times 10^{10}$ |
| $H\cdot$ | 2 | $1\cdot0 \times 10^{10}$ |
| $OH^-$ | 11–13 | $2\cdot0 \times 10^7$ |
| $O_2$ | 0·4–3 | $2\cdot6 \times 10^{10}$ |
| $H_2O_2$ | 0·4–3 | $1\cdot6 \times 10^8$ |
| $Cu^{2+}$ | 1·5 | $4\cdot2 \times 10^7$ |
| $Fe^{3+}$ | 0·4 | $1 \times 10^6$ |
| $Fe(CN)_6^{3-}$ | 1·7 | $8\cdot7 \times 10^9$ |
| $Zn^{2+}$ | 7 | $<10^5$ |
| $C_6H_6$ | 2 | $5\cdot3 \times 10^8$ |
| $HCO_2H$ | 1 | $1\cdot1 \times 10^6$ |
| $CH_3OH$ | 1 | $1\cdot6 \times 10^6$ |
| $CH_3CH_2OH$ | 1 | $1\cdot5 \times 10^7$ |
| $(CH_3)_2CHOH$ | 2 | $3\cdot9 \times 10^7$ |

**Measurement of radical yields**

Yields of free radicals produced in a variety of irradiated systems have been determined by measuring the change in concentration of an added scavenger during the radiolysis. Typical results for the radiolysis of n-hexane–iodine solutions are shown in Figure 5.5. It will be seen that the rate of disappearance of iodine is constant throughout the irradiation and is independent of the initial iodine concentration, indicating that all the radicals produced in the system are scavenged by the iodine. The overall reaction may be represented by the equations:

$$R_1H \xrightarrow{\quad\quad} R\cdot \qquad (5.49)$$

(The reaction is written for simplicity in this way but probably proceeds via ionization and excitation.)

$$R\cdot + I_2 \rightarrow RI + I\cdot \qquad (5.50)$$
$$I\cdot + I\cdot \rightarrow I_2 \qquad (5.51)$$

where $R\cdot$ represents all the radicals formed in the solution. It follows that

$$G(R) = 2G(-I_2)$$

Other scavengers that have been used include iron (III) chloride and the stable free radical diphenylpicrylhydrazyl. The reactions occurring are

$$FeCl_3 + R\cdot \rightarrow FeCl_2 + RCl \qquad (5.52)$$
$$DPPH\cdot + R\cdot \rightarrow DPPHR \qquad (5.53)$$

Typical values of the radical yields are shown in Table 5.3.

TABLE 5.3
*Radical yields determined by scavenger technique*

| Compound | $G(R)$ | | |
|---|---|---|---|
| | $I_2$ | $FeCl_3$ | DPPH |
| n-octane | 7·4 | | |
| n-hexane | 7·6 | | |
| n-pentane | 7·9 | | |
| cyclohexane | 6·2 | | 8·7 |
| benzene | 0·66 | 0·86 | 0·75 |
| dioxan | | 3·5 | 12 |
| diethyl ether | | 12·5 | 15 |

Where there is good agreement amongst the yields determined by different scavengers, the measured radical yields may be safely accepted. In some cases, however, high values are obtained and it is likely in these cases that some scavenging of excited states is also being observed. The method needs to be used with some caution.

PROBLEMS
5.1. If the radiolysis of liquid ethane is represented by reactions (5.10) and (5.11) show that the stationary concentration of ethyl radicals is given by

$$[C_2H_5\cdot]_s = \left( \frac{G(C_2H_5) \times D}{6\cdot02 \times 10^{22} \, k_{5.11}} \right)^{1/2}$$

where $D$ is the dose rate in eV cm$^{-3}$s$^{-1}$. Hence calculate the dose rate required to produce a stationary concentration of $2 \times 10^{-8}$ M in the cavity of an e.s.r. spectrometer. Assume $G(C_2H_5)=4$.

5.2. Calculate the degree of ionization of the hydroxyl radical (a) at pH 10, (b) at pH 13.

5.3. What fraction of the hydrogen atoms will react with ethanol in an oxygenated aqueous solution ($[O_2]=10^{-3}$ M) of ethanol at pH$=1$ when the concentration of ethanol is (a) $10^{-2}$ M, (b) 1 M?

5.4. A solution of $10^{-3}$ M iodine in n-hexane is irradiated. Calculate the irradiation time required to consume 10 per cent of the iodine. The dose rate in Fricke dosimeter solution is $10^{18}$ eV l$^{-1}$s$^{-1}$. The density of n-hexane is $0\cdot659$ g cm$^{-3}$.

# 6. Studies of selected systems

IN this chapter it will only be possible to consider a few of the many hundreds of systems that have been investigated. It is hoped, however, that the choice of systems will illustrate the general reactions occurring in irradiated systems so that the reader should be in a position to make intelligent guesses as to the reactions likely to occur in other systems. Undoubtedly more effort has been put into understanding the radiation chemistry of water than any other system and this will be considered first.

## Water

*Determination of primary yields*

Some of the individual species occurring in irradiated water have been discussed in Chapters 3 and 5. The overall equation for water decomposition may be represented as

$$H_2O \xrightarrow{\quad\quad} H\cdot, H^+, e_{aq}^-, \cdot OH, H_2, H_2O_2 \qquad (6.1)$$

The products $H_2$ and $H_2O_2$ are known as molecular products and $H\cdot$, $e_{aq}^-$ and $\cdot OH$ as radical products. A large part of the early work in aqueous radiation chemistry was directed to determination of the yields of these primary products.

$G_{H_2}$ can be determined directly by measuring the amount of hydrogen evolved in the radiolysis of deaerated aqueous potassium bromide. The bromide ion serves to prevent destruction of hydrogen in the reaction

$$\cdot OH + H_2 \rightarrow H_2O + H\cdot \qquad (6.2)$$

by removing hydroxyl radicals

$$Br^- + \cdot OH \rightarrow Br\cdot + OH^- \qquad (6.3)$$

The bromine atoms thus produced are unable to react with molecular hydrogen.

$G_{H_2O_2}$ can be determined directly by measuring the oxygen evolved in the radiolysis of deaerated cerium (IV) sulphate. Hydrogen peroxide reacts as follows:

$$H_2O_2 + Ce(IV) \rightarrow Ce(III) + HO_2^- + H^+ \qquad (6.4)$$

$$HO_2^- + Ce(IV) \rightarrow Ce(III) + H^+ + O_2 \qquad (6.5)$$

so that

$$G_{H_2O_2} = G(O_2)$$

The effect of scavengers on the molecular yields has been investigated. If the yields of molecular hydrogen peroxide in the presence and in the absence of scavengers are denoted as $G_{H_2O_2}$ and $(G_{H_2O_2})_0$ respectively, then the results for a typical system can be represented as in Figure 6.1.

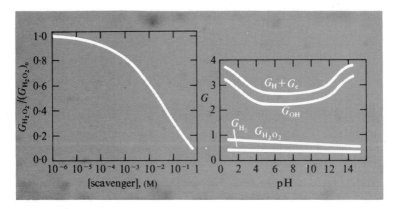

FIG. 6.1    Effect of scavenger concentration on the molecular hydrogen peroxide yield in water radiolysis

FIG. 6.2    Dependence of yields on pH for $\gamma$-radiolysis of water

It is interesting that the results of selected systems can be brought into co-incidence by multiplying the solute concentration by a normalizing factor. The dependence of $G_{H_2}$ on scavenger concentration behaves in an identical fashion. This indicates that the scavenging mechanism is similar for both products; that is the precursors of these products are scavenged by the solutes in the spur, e.g.

$$\cdot OH + \cdot OH \rightarrow H_2O_2 \tag{6.6}$$
$$\cdot OH + Br^- \rightarrow OH^- + Br \cdot \tag{6.7}$$

Clearly the normalizing factor should be related to the rate coefficient for the reaction of the scavenger with the precursor and this is found to be so.

Other scavengers however do not conform to this simple pattern. Thus it is found that iron(III) ions and copper(II) ions are much more effective in decreasing $(G_{H_2})_0$ than would be expected on the basis of their reactivities with hydrogen atoms and it is likely that these reagents interact with some precursor of the hydrogen atom. Because of the effect of reagents on the yields it is essential that the determination of the yields of all primary products be carried out at relatively low solute concentrations.

Irradiation of aqueous solutions containing the iron(II) ion is important because of the use of the Fricke dosimeter. The reactions occurring are:

$$H_2O \xrightarrow{\hspace{1cm}} H\cdot, H^+, e_{aq}^-, \cdot OH, H_2, H_2O_2 \tag{6.1}$$
$$H\cdot + O_2 \rightarrow HO_2^\cdot \tag{6.8}$$
$$e_{aq}^- + O_2 \rightarrow O_2^-\cdot \tag{6.9}$$
$$O_2^-\cdot + H^+ \rightleftharpoons HO_2^\cdot \tag{6.10}$$
$$e_{aq}^- + H^+ \rightarrow H\cdot \tag{6.11}$$
$$HO_2^\cdot + Fe^{2+} \rightarrow Fe(III) + HO_2^- \tag{6.12}$$
$$HO_2^- + H^+ \rightleftharpoons H_2O_2 \tag{6.13}$$
$$H_2O_2 + Fe^{2+} \rightarrow Fe(III) + \cdot OH + OH^- \tag{6.14}$$
$$\cdot OH + Fe^{2+} \rightarrow Fe(III) + OH^- \tag{6.15}$$

so that

$$G(Fe(III)) = 3(G_e + G_H) + 2G_{H_2O_2} + G_{OH}$$

In the presence of trace amounts of organic impurity, hydroxyl radicals will react

$$\cdot OH + RH \rightarrow R\cdot + H_2O \tag{6.16}$$
$$R\cdot + O_2 \rightarrow RO_2^\cdot \tag{6.17}$$

and the peroxy radicals thus produced may set up a chain reaction resulting in high values of $G(Fe(III))$.

$$RO_2^\cdot + Fe^{2+} + H_3O^+ \rightarrow ROOH + Fe(III) + H_2O \tag{6.18}$$
$$ROOH + Fe^{2+} \rightarrow RO^- + \cdot OH + Fe(III) \tag{6.19}$$

Reaction (6.16) may however be suppressed by the addition of chloride ion

$$\cdot OH + Cl^- \rightarrow OH^- + Cl\cdot \tag{6.20}$$

The chlorine atom is much less reactive towards organic compounds than the hydroxyl radical and will react with the iron(II) ion.

$$Cl\cdot + Fe^{2+} \rightarrow Cl^- + Fe(III) \tag{6.21}$$

It is for this reason that Fricke dosimeter solution contains $10^{-3}$ M sodium chloride.

Because of the pronounced effects that even trace amounts of impurities may have, it is important in radiation chemistry to work with materials of high purity. In aqueous studies, triply distilled water is used. Water is distilled to remove inorganic impurity. It is then redistilled from an alkaline solution of potassium permanganate which destroys organic impurity and is then finally distilled from dilute sulphuric acid.

In deaerated acidic solutions of iron(II) ion, hydrogen atoms react via

$$H\cdot \ + \ H^+ \ + \ Fe^{2+} \ \rightarrow \ Fe(III) \ + \ H_2 \tag{6.22}$$

so that

$$G(Fe(III)) \ = \ G_e \ + \ G_H \ + \ G_{OH} \ + \ 2G_{H_2O_2}$$

Knowing the value of $G_{H_2O_2}$ it is thus possible from the measurement of $G(Fe(III))$ in air-saturated and in deaerated solutions to determine $G_{OH}$ and $G_{red}$ where $G_{red}=G_e+G_H$. Although the hydrogen atom and the hydrated electron behave in a qualitatively similar fashion with a wide variety of re-agents, it is possible in a few cases to distinguish between these species. Thus dinitrogen monoxide is a specific reagent for hydrated electrons.

$$N_2O \ + \ e_{aq}^- \ \rightarrow \ N_2 \ + \ O^- \tag{6.23}$$

and

$$G(N_2) \ = \ G_e$$

Hydrated electrons react with chloroacetic acid to give chloride ion.

$$e_{aq}^- \ + \ ClCH_2CO_2H \ \rightarrow \ Cl^- \ + \ \cdot CH_2CO_2H \tag{6.24}$$

Hydrogen atoms however react mainly to give hydrogen.

$$H\cdot \ + \ ClCH_2CO_2H \ \rightarrow \ H_2 \ + \ \cdot CHClCO_2H \tag{6.25}$$

It is sometimes convenient to discuss the radiolysis of water in terms of $G(-H_2O)$. Clearly, by mass balance,

$$G(-H_2O) \ = \ G_H \ + \ G_e \ + \ 2G_{H_2}$$
$$= \ G_{OH} \ + \ 2G_{H_2O_2}$$

There is a marked effect of pH on the yields determined from the reactions of some solutes as may be seen from Figure 6.2.

The yields at low and high pH are higher than at neutral pH. This effect has been attributed to the reactions of excited water molecules with hydrogen ion and hydroxide ion.

$$H_2O^* \ + \ H^+ \ \rightarrow \ H_2^+ \ + \ \cdot OH \tag{6.26}$$
$$H_2O^* \ + \ OH^- \ \rightarrow \ e_{aq}^- \ + \ \cdot OH \tag{6.27}$$

Around neutral pH, the excited water molecule reverts to the ground state without decomposition. With other reagents such as formate ion and 2-propanol, the yields are nearly independent of pH. It may be that these solutes are able to interact with the excited species in a similar way to the hydrogen and hydroxyl ions. Although there is some controversy about the breakdown of $G_{red}$ into $G_e + G_H$, it seems likely that $G_H$ is about 0·7 and is independent of pH.

The effect of LET on the primary yields is shown in Figure 6.3. As might be

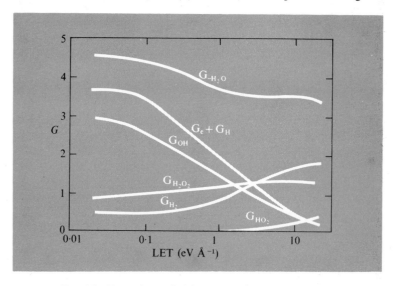

Fig. 6.3    Dependence of yields on LET for water radiolysis

expected there is an increase in the molecular products with a corresponding decrease in the radical products with increasing LET. In addition to the self-recombination of hydrogen atoms and hydroxyl radicals in the spur, there will be some recombination to give water

$$H \cdot + \cdot OH \rightarrow H_2O \qquad (6.28)$$

This results in a decrease in $G(-H_2O)$ with increasing LET. At very high LET, $HO_2 \cdot$ appears as a primary product.

In water vapour it might be expected that the extent of radical–radical recombination would be less than in liquid water and thus the yield of decomposition of water would be greater. In agreement with this it is observed that $G(-H_2O)$ increases from about 4 in liquid water to about 8 in water vapour.

*Competition kinetics*

Although the distribution of radicals may be non-uniform in an irradiated solution, it is possible in some cases to apply homogeneous kinetics. Thus consider a radical, $R\cdot$, produced in a solution where it may react with either of reagents S and X.

$$R\cdot + S \rightarrow product_1 \tag{6.29}$$
$$R\cdot + X \rightarrow product_2 \tag{6.30}$$

If these are the only reactions by which $R\cdot$ disappears, i.e. there is no bimolecular recombination of $R\cdot$,

$$R\cdot + R\cdot \nrightarrow R_2 \tag{6.31}$$

then the probability that R will react with S is given by

$$\begin{aligned} \text{probability} &= \frac{k_{6\cdot29}[R][S]}{k_{6\cdot29}[R][S] + k_{6\cdot30}[R][X]} \\ &= \frac{k_{6\cdot29}[S]}{k_{6\cdot29}[S] + k_{6\cdot30}[X]} \end{aligned}$$

This probability is independent of the concentration of $R\cdot$ and will therefore be true whether or not the distribution of R is homogeneous.

In the radiolysis of oxygen-saturated aqueous solutions of 2-propanol and methanol, hydroxyl radicals react via

$$\cdot OH + (CH_3)_2 CHOH \rightarrow (CH_3)_2\cdot COH + H_2O \tag{6.32}$$
$$\cdot OH + CH_3OH \rightarrow \cdot CH_2OH + H_2O \tag{6.33}$$

Acetone is subsequently produced as a result of the reaction

$$(CH_3)_2\cdot COH + O_2 \rightarrow (CH_3)_2 CO + HO_2\cdot \tag{6.34}$$

It follows that

$$G(CH_3COCH_3) = \frac{k_{6\cdot32}[(CH_3)_2CHOH]}{k_{6\cdot32}[(CH_3)_2 CHOH] + k_{6\cdot33}[CH_3OH]} \cdot G_{OH}$$
$$\therefore \frac{1}{G(CH_3COCH_3)} = \frac{1}{G_{OH}}\left(1 + \frac{k_{6\cdot33}[CH_3OH]}{k_{6\cdot32}[(CH_3)_2CHOH]}\right)$$

A plot of $1/G(CH_3COCH_3)$ against $[CH_3OH]/[(CH_3)_2CHOH]$ should be linear. This is found to be so as may be seen from Figure 6.4. From Figure 6.4, it will also be seen that

$$\text{intercept} = \frac{1}{G_{OH}}$$

and

$$\text{slope} = \frac{1}{G_{OH}} \cdot \frac{k_{6 \cdot 33}}{k_{6 \cdot 32}}.$$

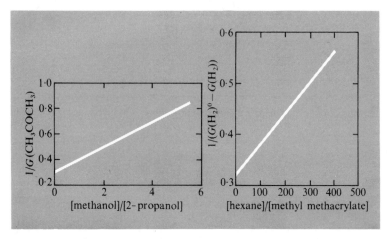

Fig. 6.4   Typical reciprocal plot for analysis of data for OH reactivities
Fig. 6.5   Kinetic analysis for determination of hydrogen atom reactivities

It is thus possible to determine $G_{OH}$ and the rate coefficient ratio. Reciprocal plots of this type have been used extensively both in aqueous and other systems to determine rate-coefficient ratios.

*Solutes of biological interest*

A considerable amount of work has been carried out on the radiolysis of aqueous solutions containing relatively simple organic molecules with a view to elucidating both the primary processes and the reactions occurring and the details of these are generally well understood. Some work has been carried out on the radiolysis of solutions of more complex compounds such as carbohydrates and proteins, and although the chemistry is necessarily more complex, such results as have been obtained are of prime importance in our understanding of the effects of radiation on living systems.

Hydrogen atoms and hydroxyl radicals react with carbohydrates via hydrogen abstraction. In evacuated solution, the radicals thus produced may initiate polymerization or may recombine with other radicals. In oxygen containing solutions, which are more relevant to living systems, the radicals undergo further oxidation and degradation. The reactions occurring can be understood to some extent by comparison with the radiolysis of ethanol which is a simpler hydroxy compound. The complex nature of the reactions may be judged by the variety of products obtained, as shown in Table 6.1.

TABLE 6.1

*Products of the radiolysis of oxygen-saturated aqueous*
*D-glucose*

| No. of C atoms in chain | Compound |
|---|---|
| 6 | D-glucuronic acid |
|   | D-gluconic acid |
| 5 | D-arabinose |
| 4 | D-erythrose |
| 3 | D-glyceraldehyde |
|   | dihydroxy acetone |
| 2 | glyoxal |
|   | glycollic aldehyde |

The radiation chemistry of aqueous solutions of the simple amino acids glycine and alanine has been extensively investigated in the hope that these would serve as model systems for the understanding of protein radiolysis. The chemistry of even these simple systems is however relatively complex. In oxygen-saturated aqueous solutions of glycine, the products arise largely from hydroxyl radical attack on the glycine.

$$\cdot OH + NH_2CH_2CO_2H \rightarrow NH_2 \cdot CHCO_2H + H_2O \quad (6.35)$$

$$NH_2 \cdot CHCO_2H + O_2 \rightarrow HN=CHCO_2H + HO_2 \cdot \quad (6.36)$$

$$HN=CHCO_2H + H_2O \begin{cases} \nearrow \begin{array}{c} CHO \\ | \\ CO_2H \end{array} + NH_3 \quad (6.37) \\ \searrow CH_2O + NH_3 + CO_2 \quad (6.38) \end{cases}$$

Methylamine is also formed in 1 M glycine and may arise from direct excitation of glycine.

$$NH_2CH_2CO_2H \longrightarrow\!\!\!\!\wedge\!\!\!\!\wedge\!\!\!\!\rightarrow CH_3NH_2 + CO_2 \quad (6.39)$$

### Dinitrogen monoxide

The radiolysis of gaseous dinitrogen monoxide is of some interest since this system is often used as a dosimeter in the study of gaseous systems. $G(N_2)$ is 10·2 and is independent of the pressure of nitrous oxide in the range 200–600 mm Hg. It is not very dependent on LET. After the radiolysis, unreacted dinitrogen monoxide and any other oxides of nitrogen can be frozen out and the residual nitrogen determined by conventional gasometric techniques. $G(-N_2O)$ is 12 and since $W$ for dinitrogen monoxide is about 30 eV, the ionic yield $M/N$ is 4. The high value of the ionic yield indicates that either a short chain reaction is occurring or that reactions of excited molecules are important. The detailed reaction mechanism has not yet been fully elucidated but it appears that the primary processes of ionization and excitation

$$N_2O \xrightarrow{\quad\rightsquigarrow\quad} N_2O^+ + e^- \qquad (6.40)$$
$$N_2O \xrightarrow{\quad\rightsquigarrow\quad} N_2O^* \qquad (6.41)$$

are followed by neutralization and dissociative electron capture.

$$N_2O^+ + e^- \rightarrow N_2O^* \qquad (6.42)$$
$$N_2O + e^- \rightarrow N_2 + O\cdot^- \qquad (6.43)$$

There is some mass spectrometric evidence for positive-ion decomposition.

$$N_2O^+ \rightarrow NO^+ + N\cdot \qquad (6.44)$$

Excited dinitrogen monoxide decomposes either by

$$N_2O^* \rightarrow N\cdot + NO\cdot \qquad (6.45)$$

or

$$N_2O^* \rightarrow N_2 + O\cdot \qquad (6.46)$$

The resulting atoms of oxygen and nitrogen can undergo a variety of reactions.

$$O\cdot + O\cdot + M \rightarrow O_2 + M \qquad (6.47)^\dagger$$
$$N\cdot + N\cdot + M \rightarrow N_2 + M \qquad (6.48)$$
$$N\cdot + NO\cdot \rightarrow N_2 + O\cdot \qquad (6.49)$$
$$O\cdot + N_2O \rightarrow O_2 + N_2 \qquad (6.50)$$

**Oxygen**

The radiolysis of oxygen is of interest since such reactions, initiated by cosmic radiation, might be expected to be important in the earth's upper atmosphere. It has also been hoped that the radiation induced oxidation of nitrogen–oxygen mixtures would provide a suitable alternative process for the fixation of nitrogen.

Ozone is the product of irradiation of oxygen at 293 K and 760 mm Hg. Pulse radiolysis, using high-energy electrons, indicates a value of $G(O_3)$ of 13·8

† Recombination of two atoms to form a molecule proceeds efficiently only in the presence of a third body, M, to remove the energy of bond formation.

though values of 10 and 6 have been obtained for the continuous irradiation by $^{60}Co$ $\gamma$-rays and $\alpha$-particles respectively. The reaction is complicated by the ease with which ozone reacts either with impurities or with the radical products of the radiolysis. The principal reactions leading to the production of ozone are:

$$O_2 \longrightarrow\!\wedge\!\wedge\!\rightarrow O_2^+ + e^- \qquad (6.51)$$

$$O_2 \longrightarrow\!\wedge\!\wedge\!\rightarrow 2O\cdot \qquad (6.52)$$

$$O_2 \longrightarrow\!\wedge\!\wedge\!\rightarrow O_2^* \qquad (6.53)$$

$$O_2 + {}'e^- \rightarrow O_2^-\cdot \qquad (6.54)$$

$$O_2^+ + O_2^-\cdot \rightarrow O_2^* + O_2 \qquad (6.55)$$

$$O_2^* + O_2 \rightarrow O_3 + O\cdot \qquad (6.56)$$

$$O\cdot + O_2 + M \rightarrow O_3 + M \qquad (6.57)$$

Decomposition of ozone proceeds via the reactions

$$O\cdot + O_3 \rightarrow 2O_2 \qquad (6.58)$$

$$O_3 + e^- \rightarrow O_3^- \qquad (6.59)$$

and will be more important as the concentration of ozone increases.

It is possible that the germicidal effect of ionizing radiation is in part due to the production of ozone in irradiated air.

In mixtures of oxygen and nitrogen, in addition to reactions (6.51)–(6.59), the following reactions can take place:

$$N_2 \longrightarrow\!\wedge\!\wedge\!\rightarrow N_2^+ + e^- \qquad (6.60)$$

$$N_2 \longrightarrow\!\wedge\!\wedge\!\rightarrow 2N\cdot \qquad (6.61)$$

$$N\cdot + O_2 \rightarrow NO\cdot + O\cdot \qquad (6.62)$$

$$2NO\cdot + O_2 \rightarrow 2NO_2 \qquad (6.63)$$

$$NO\cdot + N\cdot \rightarrow N_2 + O\cdot \qquad (6.64)$$

$$N\cdot + \cdot NO_2 \rightarrow N_2O + O\cdot \qquad (6.65)$$

The ionization potential of nitrogen monoxide is relatively low and positively charged ions might be expected to undergo charge transfer with nitrogen monoxide. (All positively charged ions for convenience will be denoted by $M^+$).

$$M^+ + NO\cdot \rightarrow NO^+ + M \qquad (6.66)$$

$$NO^+ + e^- \rightarrow N\cdot + O\cdot \qquad (6.67)$$

$$NO^+ + e^- \rightarrow NO\cdot + h\nu \qquad (6.68)$$

The principal products are nitrogen dioxide and dinitrogen monoxide. The yields depend on the composition and pressure of the gas mixture but are generally low with $G(-N_2)$ about 5.

## Hydrocarbons

Hydrocarbons have been the most widely investigated class of organic compounds. This attention has, in part, been directed by the interest of oil companies in the possible modification of hydrocarbon structures by radiation.

The principal gaseous product in the radiolysis of saturated hydrocarbons is hydrogen, which arises by both molecular and free radical processes. That this is so is evidenced by the observation that only a part of the hydrogen yield is eliminated by the addition of scavengers known to react with hydrogen atoms. In n-hexane, for example, the reactions occurring are

$$C_6H_{14} \xrightarrow{G_1} C_6H_{13}\cdot + H\cdot \tag{6.69}$$

$$C_6H_{14} \xrightarrow{G_2} C_6H_{12} + H_2 \tag{6.70}$$

$$H\cdot + C_6H_{14} \rightarrow C_6H_{13}\cdot + H_2 \tag{6.71}$$

If a solute S is added which may react with hydrogen atoms, either by addition or by hydrogen abstraction,

$$S + H\cdot \rightarrow SH\cdot \tag{6.72}$$

$$S + H\cdot \rightarrow product + H_2 \tag{6.73}$$

then the decrease in hydrogen yield, $\Delta G(H_2)$, is given by

$$\Delta G(H_2) = G_1 \left( \frac{k_{6.72}[S]}{k_{6.72}[S] + k_{6.73}[S] + k_{6.71}[C_6H_{14}]} \right)$$

(It is assumed that the concentration of solute is kept sufficiently low, about 2 per cent, so that the yields of the primary processes (6.69) and (6.70) are unaffected by it and the subsequent reactions of SH· radicals do not yield hydrogen).

It follows, therefore, that

$$\frac{1}{(G(H_2)^0 - G(H_2))} = \frac{1}{G_1}\left(1 + \frac{k_{6.73}}{k_{6.72}}\right) + \frac{1}{G_1} \cdot \frac{[C_6H_{14}]}{[S]} \cdot \frac{k_{6.71}}{k_{6.72}}$$

where $G(H_2)^0$ and $G(H_2)$ are the hydrogen yields in the absence and presence of solute, respectively. Clearly $G(H_2)^0 = G_1 + G_2$. If a solute is used which reacts only by addition with hydrogen atoms, e.g. an unsaturated compound, then $k_{6.73}$ is zero and it is possible to determine $G_1$ and $k_{6.71}/k_{6.72}$. Results for a typical system are shown in Figure 6.5 and are in good agreement with this equation. Under these conditions, it follows that

$$\text{intercept of line in Figure 6.5} = \frac{1}{G_1}$$

and

$$\text{slope of line in Figure 6.5} = \frac{1}{G_1} \cdot \frac{k_{6.71}}{k_{6.72}}.$$

Values of $k_{6.72}$ are known for selected reactants from gas phase thermal oxidation studies and so it is possible to determine an absolute value for $k_{6.71}$. Thus it has been established that $G_1$ and $G_2$ are 3·16 and 2·12 respectively for high-energy electrons and that $k_{6.71}$ is $4·9 \times 10^6$ $M^{-1}s^{-1}$ at 296 K. Using these values, it is possible to determine values of $k_{6.72}$ and $k_{6.73}$ for a wide variety of organic reactants. Some selected values are given in Table 6.2.

TABLE 6.2

*Rate coefficients for hydrogen atom reactions at 296 K*

| Reactant | Hydrogen addition $10^{-7}k_{6.72}$ $(M^{-1}s^{-1})$ | Hydrogen abstraction $10^{-7}k_{6.73}$ $(M^{-1}s^{-1})$ |
|---|---|---|
| n-hexane | 0 | 0·49 |
| l-hexene | 79 | 24 |
| benzene | 18 | <1 |
| anthracene | 310 | 0 |
| DPPH | 1050 | 0 |
| acetic acid | 61 | 15 |
| acetone | 32 | 16 |

The organic products of the radiolysis of hydrocarbons are usually complex. Information on the breakdown pattern of a hydrocarbon can be obtained by irradiating a solution of iodine in the hydrocarbon and identifying the alkyl iodides produced. This can be done either by using radioactive iodine and identifying the iodides by isotopic dilution analysis or by using inactive iodine and gas–liquid chromatography to identify the iodides. Results obtained for n-hexane are shown in Table 6.3.

TABLE 6.3

*Alkyl iodides formed in the irradiation of $C_6H_{14}$–$I_2$ solutions*

| Alkyl iodide | G(RI) |
|---|---|
| n-hexyl iodide | 0·70 |
| s-hexyl iodide | 2·60 |
| n-butyl iodide | 0·30 |
| n-propyl iodide | 0·70 |
| ethyl iodide | 0·40 |
| methyl iodide | 0·10 |

The results are in agreement with the e.s.r. observations (p. 47) that the radicals are principally produced as a result of C−H rather than C−C bond fission. Similar conclusions have been reached by using oxygen as a scavenger and identifying the alkyl peroxides produced. Aromatic compounds are much more radiation inert, i.e. the decomposition yields are smaller, than aliphatic compounds. This is attributed to the ability of the conjugated ring system to degrade excitation energy without decomposition. For this reason polyphenyls are sometimes used as coolants and moderators in nuclear reactors. Polyphenyl ethers are similarly used as radiation-resistant lubricants.

### Polymerization of monomers

In 1874 the formation of an inert solid, which we now know to be a polymeric material, when acetylene was exposed to an electric discharge, was observed. Since that first observation, the polymerization of a wide variety of monomers by ionizing radiation has been investigated. As in conventional polymerization, two different mechanisms can be distinguished—ionic and free-radical polymerization.

#### Ionic polymerization

Examples of monomers that undergo typical cationic polymerization are isobutene and cyclopentadiene. Such systems are characterized by the fact that polymerization is markedly suppressed by substrates which can readily

$$RH^+ + NH_3 \rightarrow R\cdot + NH_4^+ \tag{6.74}$$

react with cations, such as ammonia, amines, and water, but is much less markedly suppressed by free-radical scavengers such as oxygen and DPPH. In some systems, however, oxygen may suppress the polymerization by capturing secondary electrons.

$$O_2 + e^- \rightarrow O_2^-\cdot \tag{6.75}$$

Anionic polymerization has been demonstrated in nitroethylene.

#### Free-radical polymerization

Monomers that have been investigated and shown to polymerize by a free-radical mechanism include ethylene, vinyl halides and acrylamide. The polymerization of acrylamide in aqueous solution was one of the earliest indications of the existence of free radicals in water radiolysis.

The basic reactions occurring in an irradiated system may be represented as:

$$\text{Initiation M} \quad \xrightarrow{\quad} \quad \text{R·} \qquad (6.76)$$

$$\text{Propagation R·} + \text{M} \quad \rightarrow \quad \text{P} + \text{R·} \qquad (6.77)$$

$$\text{Termination R·} + \text{R·} \quad \rightarrow \quad \text{R}_2 \qquad (6.78)$$

By convention R is used to represent all radicals occurring in the monomer M and P represents the growing polymer chains. A steady state will be set up such that the rates of production and disappearance of radicals are the same.

i.e.
$$\frac{d[\text{R·}]}{dt} = 0 = R_i - k_t[\text{R·}]^2$$

where $R_i$ is the rate of production of radicals.

The overall reaction rate is equal to the rate of disappearance of monomer and, for long chains, this is given by

$$-\frac{d[\text{M}]}{dt} = k_p[\text{R·}][\text{M}]$$

Since
$$[\text{R·}] = \frac{R_i^{1/2}}{k_t^{1/2}}$$

we have

$$-\frac{d[\text{M}]}{dt} = \frac{k_p}{k_t^{1/2}} R_i^{1/2}[\text{M}]$$

where $k_p$ and $k_t$ are the rate coefficients for the propagation and termination steps respectively. This expression is analogous to that derived for systems with added initiators.

If $D$ is the dose rate in eV $l^{-1}s^{-1}$ then $R_i$ is given by

$$R_i = \frac{D}{100} \cdot \frac{G(\text{R})}{N}$$

where $N$ is Avogadro's number and $G(\text{R})$ is the total radical yield for the irradiated monomer. This can usually be determined by measuring the consumption of free-radical scavengers—e.g. iron(III) chloride or iodine—in the irradiated system, and so it is possible to evaluate $k_p/k_t^{1/2}$. There is good agreement between the values obtained from radiation studies and from other initiated systems. The yields of initiating species, whether free-radical or ionic, are usually in the range 1–10. However, because of the chain nature of the reactions, $G(-\text{monomer})$ may be as high as $10^5$.

There is seldom any advantage of the radiation-initiated system over other systems. The radiation-induced polymerization does not need an added initiator for polymerization to take place and so may yield a somewhat purer product. Some fluorinated monomers such as perfluoropropylene and perfluoroacrylonitrile are polymerized by $\gamma$-rays but not readily by any other means.

There has recently been considerable interest in the radiation-induced polymerization of solid monomers. It has been shown that acrylonitrile and acrylamide polymerize by a free-radical mechanism but trioxane by an ionic mechanism.

### Irradiation of polymers

The two principal changes occurring when a polymer is irradiated are degradation and crosslinking. In addition some decomposition to give gaseous products, usually of small molecules such as hydrogen, methane, carbon monoxide, or carbon dioxide, may be observed. As a general rule, polymers of the type $(-CH_2C(CH_3)R-)_n$ undergo degradation and those of the type $(-CH_2CHR-)_n$ undergo crosslinking.

### *Degradation*

Degradation occurs as a result of main chain fission;

$$-A-A-A-A- \quad \leadsto \quad -A-A\cdot \; + \; \cdot A-A- \qquad (6.79)$$

As indicated in Chapter 5, this chain fission probably arises as a result of ionizations and excitations. Degradation is usually promoted by the presence of oxygen which acts by forming peroxides.

$$-A-A\cdot \; + \; O_2 \quad \rightarrow \quad -A-A-O_2\cdot \qquad (6.80)$$

and thus effectively prevents any recombination of the radical ends of the chains. As would be expected from the general rule, polystyrene and poly-(methyl methacrylate) are examples of polymers that undergo degradation. Typically, the yield of main-chain breaks, $G(\text{scission})$, ranges from 16, for poly-($\alpha$-methyl cellulose), to $0.3$ for poly-($\alpha$-methyl styrene).

Hydrogen is also produced in the radiolysis of poly-(methyl methacrylate). At doses below 50 Mrad, there is no visible formation of bubbles in the polymer, but if the polymer is heated above the second-order transition point, it expands forming a white brittle material which may be several times its original size, consisting of a mass of bubbles.

Degradation results in a loss of plasticity and structural strength and is usually, therefore, an undesirable process. By incorporating a few per cent of aniline or a substituted thiourea in the polymer, the extent of degradation can be reduced. These protective agents probably function by an energy-transfer mechanism.

*Crosslinking*

Crosslinking occurs when radical sites in adjacent molecules combine.

$$\begin{array}{ccc}
\sim\!\!\sim C \sim\!\!\sim & & \sim\!\!\sim C \sim\!\!\sim \\
\cdot & & \mid \\
\cdot & \rightarrow & \mid \\
\sim\!\!\sim C \sim\!\!\sim & & \sim\!\!\sim C \sim\!\!\sim
\end{array} \qquad (6.81)$$

Such radical sites may be produced by two processes.
1. Loss of a hydrogen atom from the irradiated polymer

$$\sim\!CH_2\!\sim \quad \xrightarrow{\sim\!\sim\!\sim} \quad \sim\!\cdot CH\!\sim \; + \; H\cdot \qquad (6.82)$$

2. Addition of a hydrogen atom to an unsaturated group

$$\sim\!CH\!=\!CH\!\sim \; + \; H\cdot \quad \rightarrow \quad \sim\!CH_2\!-\!\cdot CH\!\sim \qquad (6.83)$$

The radical site may be localized or it may migrate along the chain until it encounters a radical site on another molecule when crosslinking will occur.

$G$(crosslinking) in polyethylene varies from 1 at 173 K to 3 at 296 K. Crosslinking results in increased tensile strength and greater insolubility of the polymer. Since these are generally desirable properties there is considerable industrial interest in radiation-induced crosslinking. Thus whereas unirradiated polythene softens in the range 360–370 K and melts at about 390 K, it can be taken up to 520 K without losing its shape if given a dose of $2 \times 10^6$ rads.

**Graft copolymerization**

If a radical site is produced in a polymer A, which is then exposed to a monomer B, graft copolymerization of B on A may occur. This gives rise to a structure of the form

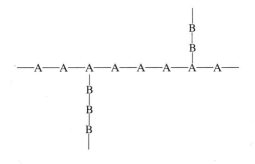

Grafting in this way is particularly useful if it is desired to modify the surface properties of a polymer without significantly affecting the bulk properties of the material. Polyethylene, because of its non-polar surface, is unsuitable as a printing medium. Polar substituents may be grafted on thus making printing of the resultant material possible.

The chief ways of producing a graft copolymer are:

1. Irradiation of the polymer in the presence of the monomer. In this case direct polymerization of the monomer is a troublesome side effect which can be minimized by using an aromatic monomer such as styrene for which $G$(initiation) is low.
2. If the radical sites are relatively stable, the monomer may be added to the polymer after the irradiation.
3. The polymer is irradiated in air which results in the production of peroxides. The monomer is then added and the mixture is heated. Decomposition of the peroxides leads to the production of radical sites on which the monomer grafts.

## PROBLEMS

6.1. Suggest a reaction mechanism for the radiolysis of deaerated aqueous solutions of Ce(IV), and hence derive an expression for the reduction yield, $G\{-Ce(IV)\}$.

6.2. An equimolar mixture of nitrogen and oxygen at a total pressure of one atmosphere at 298 K is circulated through a 100 litre reaction vessel in an irradiation facility at a dose rate of $10^5$ rad per hour and the oxides of nitrogen produced are condensed out. Calculate the consumption of nitrogen per day if $G(-N_2)$ for this system is 5. Is this likely to provide a suitable basis for an industrial process for the fixation of nitrogen?

6.3. Suggest reaction mechanisms to account for the radiolysis of aqueous, oxygen-saturated solutions of D-glucose (see Table 6.1).

6.4. Calculate $G(H_2)$ and $G(-DPPH)$ for the high-energy electron radiolysis of a $5 \times 10^{-3}$ M solution of DPPH in n-hexane. (see p. 69).
Density of n-hexane is 0.659 g cm$^{-3}$.

# 7. Applications of radiation chemistry

## General considerations

IN the 1950s and early 1960s there was considerable hope that economically viable industrial processes based on radiation chemistry might be established. There are good reasons in favour of such processes.

1. Large quantities of radioactive material are now available as by-products of nuclear fission from atomic reactors. At present these materials are buried at safe distances underground and their energy is thus spent in a useless fashion. It would clearly be desirable to harness the activity of these products in a useful way.

2. Radiation-induced reactions can be carried out at lower temperatures than those that depend on thermal initiation. This may eliminate unwanted side reactions.

3. Many free-radical processes require the addition of a chemical initiator whose presence may lead to contamination of the end product. Radiation-induced reactions need no chemical initiator and so a purer product may be obtained.

4. Photoinduced reactions must necessarily be carried out in vessels transparent to the radiation, e.g. glass or silica vessels. Steel vessels can be used for $\gamma$-initiated reactions because of the very high penetrating power of the $\gamma$-radiation.

There are however disadvantages.

1. Although the fission products are available relatively cheaply, the capital cost of building suitable protective housing for these is relatively high. Especially stringent safety precautions need to be taken.

2. The yield of primary products in most irradiated systems is low; $G = 1$–$10$. Useful systems will therefore be limited either to those in which it is difficult to prepare the product by any other method or to those in which the primary products initiate chain reactions, so that much higher product yields, $G = 10^3$–$10^5$, can be obtained.

It must be admitted that the optimism of the 1950s has not been justified and there are, as yet, only a very limited number of useful industrial processes based on radiation chemistry. Some of these are surveyed in the remainder of this chapter.

## Production of ethyl bromide

Ethylene reacts with hydrogen bromide when irradiated with $\gamma$-rays at room temperature.

$$C_2H_4 + HBr \xrightarrow{\quad\quad} C_2H_5Br \tag{7.1}$$

The reaction is a chain process with $G(C_2H_5Br) \approx 10^5$. A very pure product ($> 99.5$ per cent purity) is obtained.

### Production of gammexane

The reaction of chlorine with benzene to give gammexane ($\gamma$-benzene hexachloride) is a chain reaction that can be conveniently initiated by ionizing radiation.

$$C_6H_6 + 3Cl_2 \xrightarrow{\quad\quad} C_6H_6Cl_6 \tag{7.2}$$

$G$(Gammexane) is about $10^5$. Gammexane is widely used as an insecticide.

### Production of detergents

The sulphoxidation of alkanes can be initiated in high yields by ionizing radiation.

$$RH + SO_2 + \tfrac{1}{2}O_2 \xrightarrow{\quad\quad} RSO_3H \tag{7.3}$$

At present aryl linear-alkyl sulphonates can be produced more cheaply by non-radiation processes. These are, however, less easily biodegradable than the linear-alkyl sulphonates that can be produced by the radiation process. It is possible that with the increase in concern over environmental pollution, the radiation process may become more attractive.

### Nitrogen fixation

It has been shown that nitrogen and oxygen can be made to combine directly in a reactor core.

$$N_2 + O_2 \xrightarrow{\quad\quad} 2NO\cdot \tag{7.4}$$

$$2N_2 + O_2 \xrightarrow{\quad\quad} 2N_2O \tag{7.5}$$

In this way fission energy is converted directly to chemical energy without the need to build additional housing for the source. There are, however, problems of radioactive contamination of the products. The yields are relatively low, $G$(oxides of nitrogen) is about 4, and, as yet, the process is not economically viable.

### Polymerization

The production of graft copolymers has been discussed in Chapter 6 and some industrial processes based on this have been established. Polyethylene has only poor adhesive properties and cannot therefore be used for printing.

However if methylacrylic acid is grafted on, it is possible to print on the resultant product.

By coating wood with a monomer which is then polymerized, the porosity of the wood is considerably reduced. The resultant product is expensive.

The curing of paints is a polymerization process and conventionally has been carried out slowly in curing ovens. Certain paints can be rapidly cured by irradiation. For this purpose high-intensity electron sources are used. The rapid throughput of material is obviously an attractive feature of this process.

## Sterilization

Undoubtedly one of the most useful applications of radiation sources has been in sterilization, particularly of medical equipment. Sterilization depends on the fact that ionizing radiation is lethal to all forms of life. In general, the lethal dose is less the higher the form of life. Lethal doses are usually expressed as LD 50/30 days, i.e. the dose of radiation which kills 50 per cent of the population within thirty days. The lethal dose for man is about 400 R, for insects, $10^4$ R, and for bacteria, $5 \times 10^5$ R. (The maximum permissible dose rate for safe working for man is 0·5 mR per hour).

Since very high doses are required for sterilization of medical equipment, in practice large (about $10^5$ Ci) $^{60}$Co sources are used. Radiation sterilization is particularly useful in that it can be used on heat-sensitive materials. Thus disposable syringes made largely of plastic can be sealed in plastic bags and sterilized. Pharmaceutical products, which might lose their activity if heated, can also be conveniently sterilized.

The lethal effect of radiation is also made use of in the destruction of pests in grain and wool.

## Food preservation

The micro-organisms responsible for the deterioration of foodstuffs can be killed by suitably large doses of radiation. At the same time, however, the radiation produces chemical changes in the foodstuff which may result in the production of unacceptable flavours or aromas. As a compromise, the foodstuff may be given a lower dose of radiation which, while it does not completely sterilize the product, results in an increased lifetime. Thus the sprouting of potatoes may be inhibited by a dose of $10^4$ R. Radiation-preserved bacon has been submitted to extensive testing in the U.S.A. There are still, however, several problems to be overcome before irradiation can become widely acceptable as a method of food preservation. In particular, the destruction of vitamins by the radiation is especially undesirable.

## Radiobiology

Knowledge of the reactions occurring in the radiolysis of simpler chemical systems is leading to an understanding of the more complex reactions occurring in living systems. Although it is to be hoped that mankind will never need to face the horror of a nuclear war, it would clearly be desirable if a reagent could be discovered which, when taken by human beings, would protect them against the lethal effects of ionizing radiation. Such a reagent has not yet been discovered. However, considering the tremendous developments that have been made in our understanding of radiation chemistry in the past thirty years, it may be hoped that such a task is not beyond the bounds of possibility.

# Answers to problems

1.1.  $\lambda = \dfrac{0 \cdot 69}{30 \times 365 \times 24 \times 60 \times 60} = 7 \cdot 29 \times 10^{-10} \, \text{s}^{-1}$

1000 Ci $= 3 \cdot 7 \times 10^{13}$ disintegrations $\text{s}^{-1}$

Rate of disintegration $= - \dfrac{\mathrm{d}N}{\mathrm{d}t} = \lambda N$

i.e. $3 \cdot 7 \times 10^{13} = 7 \cdot 29 \times 10^{-10} \, N$

$N = 5 \cdot 07 \times 10^{22}$

but molecular weight $= 172 \cdot 5$

$\therefore$  weight $= 145$ g.

1.2.  $^{60}\text{Co}$  $t_{1/2} = 5 \cdot 27$ yr

$\therefore \quad \lambda = \dfrac{0 \cdot 69}{5 \cdot 27} = 0 \cdot 131 \, \text{yr}^{-1}$

After 1 year, $\log_e \left( \dfrac{\text{activity}}{500} \right) = - \, 0 \cdot 131 \times 1$

activity $= 438$ Ci.

After 10 years, $\log_e \left( \dfrac{\text{activity}}{500} \right) = - \, 0 \cdot 131 \times 10$

activity $= 135$ Ci.

For $^{137}\text{Cs}$  $\lambda = \dfrac{0 \cdot 69}{30} = 0 \cdot 023 \, \text{yr}^{-1}$

After 1 year, $\log_e \left( \dfrac{\text{activity}}{500} \right) = -0 \cdot 023 \times 1$

activity $= 489$ Ci.

After 10 years, $\log_e \left( \dfrac{\text{activity}}{500} \right) = -0 \cdot 023 \times 10$

activity $= 397$ Ci.

There is a very marked reduction of the $^{60}\text{Co}$ activity especially after 10 years but a much smaller reduction in the activity of the $^{137}\text{Cs}$.

1.3.  For $^{3}\text{H}$  $t_{1/2} = 12 \cdot 26$ yr

$\lambda = \dfrac{0 \cdot 69}{12 \cdot 26} = 0 \cdot 0563 \, \text{yr}^{-1}$

Since the time is short compared with the half life, we can calculate the approximate disintegration rate by

$- \dfrac{\mathrm{d}N}{\mathrm{d}t} = \lambda N$

i.e. $- \dfrac{\mathrm{d}N}{\mathrm{d}t} = 0 \cdot 0563 \times \dfrac{5}{100} \times \dfrac{10}{22} \times 2 \times 6 \cdot 02 \times 10^{23}$

$= 1 \cdot 54 \times 10^{21} \, \text{yr}^{-1}$

i.e. $1 \cdot 54 \times 10^{21} \, \beta$-particles are emitted per year.

1.4  For $^{222}\text{Rn}$  $t_{1/2} = 3 \cdot 83$ d

$\lambda = \dfrac{0 \cdot 69}{3 \cdot 83} = 0 \cdot 180 \, \text{d}^{-1}$

The time is no longer short compared with the half life and we can not use the approximation as in question 3. The number of disintegrations $\Delta N$ is given by

$$\Delta N = N_0 - N, \text{ but } \frac{N_0 - N}{N_0} = 1 - e^{-\lambda t}$$

$$\therefore \quad \Delta N = N_0(1 - e^{-\lambda t})$$

$$N_0 = \frac{0 \cdot 1 \times 10^{-3}}{22 \cdot 4} \times 6 \cdot 02 \times 10^{23} = 2 \cdot 69 \times 10^{18}$$

$$\therefore \quad \Delta N = 4 \cdot 43 \times 10^{17}$$

but energy of $\alpha$-particle is 5·49 MeV

$\quad \therefore \quad$ Energy absorbed $= 2 \cdot 43 \times 10^{18}$ MeV day$^{-1}$.

2.1(a) It is not necessary to use the densities to evaluate the molarities explicitly. Thus, let volume of the solution be $V$ litres

Molarity of benzene $= \dfrac{10}{78} \times \dfrac{1}{V}$

Molarity of acetone $= \dfrac{100}{46} \times \dfrac{1}{V}$

$\quad \therefore \quad$ Fraction of u.v. light absorbed by benzene

$$= \frac{170 \times 10/78 \times 1/V}{(170 \times 10/78 \times 1/V) + (18 \times 100/46 \times 1/V)}$$

$$= 0 \cdot 358$$

$\quad \therefore \quad$ Amount of light absorbed by benzene $= 35 \cdot 8$ per cent.

(b) Absorption of 1 MeV electrons will be proportional to electron fraction.

1 molecule benzene contains 42 electrons

1 molecule acetone contains 32 electrons

$\quad \therefore \quad$ Electron fraction of benzene

$$= \frac{10/78 \times 42}{10/78 \times 42 + 100/46 \times 32} = 0 \cdot 072$$

$\quad \therefore \quad$ Amount of energy absorbed by benzene $= 7 \cdot 2$ percent

2.2. Linear absorbtion coefficient, $\mu$, is defined by

$I = I_0 e^{-\mu x}$

For lead $\quad \frac{1}{2}I_0 = I_0 e^{-\mu.0 \cdot 6}$

$$\mu = \frac{\log_e 2}{0 \cdot 6} = 1 \cdot 15 \text{ cm}^{-1}$$

$\quad \therefore \quad$ mass absorption coefficient $= \dfrac{1 \cdot 15}{11 \cdot 3} = 0 \cdot 102 \text{ cm}^2 \text{ g}^{-1}$.

For water $\quad \mu = \dfrac{\log_e 2}{11} = 0 \cdot 063 \text{ cm}^{-1}$

$\quad \therefore \quad$ mass absorption coefficient $= \dfrac{0 \cdot 063}{0 \cdot 998} = 0 \cdot 063 \text{ cm}^2 \text{ g}^{-1}$.

Reduction in intensity $10^4 \quad \rightarrow \quad 1$ (i.e. $1 = 10^4 e^{-\mu x}$)

$$\therefore \quad x = \frac{4 \times 2 \cdot 303}{\mu}$$

For lead, $x = 8 \cdot 01$ cm.

For water, $x = 146 \cdot 2$ cm.

2.3.   1 rad $= 10^{-5}$ J g$^{-1}$

$\therefore$   Energy deposited $= 10 \times 10 \times 60 \times 10^{-5}$ J g$^{-1}$
$= 6 \times 10^{-2}$ J g$^{-1}$

$\therefore$   Temperature rise $= \dfrac{6 \times 10^{-2}}{0 \cdot 126} = 0 \cdot 48$ K

2.4.   Molecule of cyclohexane contains 48 electrons.

Molar volume of cyclohexane $= \dfrac{84}{0 \cdot 779}$ cm$^3$

$\therefore$   Electron density $= \dfrac{48}{84/0 \cdot 779}$
$= 0 \cdot 445$ g. electrons cm$^{-3}$

Electron density of Fricke dosimeter $= 0 \cdot 567$ g. electrons cm$^{-3}$

$\therefore$   dose rate in cyclohexane $= \dfrac{0 \cdot 445}{0 \cdot 567} \times 10^{19}$
$= 7 \cdot 85 \times 10^{18}$ eV $1^{-1}$ s$^{-1}$.

2.5.   Range $= 2 \cdot 3 \times 10^8$ Å

Mean LET $= \dfrac{4 \times 10^6}{2 \cdot 3 \times 10^8} = 1 \cdot 74 \times 10^{-2}$ eV Å$^{-1}$

From Table 2.1, mean LET for $5 \cdot 3$ MeV $\alpha$ particles in water is $13 \cdot 6$ eV Å$^{-1}$ i. e. LET increases with density.

3.1.   $k(e_{aq}^- + O_2) = 1 \cdot 9 \times 10^{10}$ M$^{-1}$ s$^{-1}$
$k(e_{aq}^- + H_2O_2) = 1 \cdot 3 \times 10^{10}$ M$^{-1}$ s$^{-1}$

$\therefore$   $\dfrac{99}{1} = \dfrac{1 \cdot 3 \times 10^{10} \,[H_2O_2]}{1 \cdot 9 \times 10^{10} \times 2 \times 10^{-4}}$

$[H_2O_2] = 2 \cdot 89 \times 10^{-2}$ M

3.2.   At $10^{-8}$ M, self recombination of $e_{aq}^-$ is negligible and $e_{aq}^-$ disappears at pH 3 by reaction with $H_3O^+$.

$k(e_{aq}^- + H_3O^+) = 2 \cdot 06 \times 10^{10}$ M$^{-1}$ s$^{-1}$

Pseudo first-order rate coefficient, $k'$, is given by

$k' = k[H_3O^+]$,

and $t_{1/2} = \dfrac{\log_e 2}{k'}$

$= \dfrac{0 \cdot 69}{2 \cdot 06 \times 10^{10} \times 10^{-3}}$
$= 3 \cdot 35 \times 10^{-8}$ s.

3.3.   $1 \cdot 6 \times 10^1 \,[e_{aq}^-]\,[H_2O] = 5 \times 10^9 \,[e_{aq}^-]^2$

$\therefore$   $[e_{aq}^-] = \dfrac{1 \cdot 6 \times 10^1 \times 55}{5 \times 10^9}$
$= 1 \cdot 76 \times 10^{-7}$ M.

4.1   $t_{1/2} = \dfrac{\log_e 2}{k}$

$\therefore$   $k = \dfrac{0 \cdot 69}{2 \cdot 3 \times 10^{-9}} = 3 \times 10^8$ s$^{-1}$

Fraction of benzene triplets reacting with anthracene is given by

$$\frac{3\cdot9 \times 10^{10} \times 5 \times 10^{-3}}{(3\cdot9 \times 10^{10} \times 5 \times 10^{-3}) + 3 \times 10^{8}} = 0\cdot394$$

$$\therefore \quad G(\text{anthracene}_T) = 4\cdot2 \times 0\cdot394 = 1\cdot65.$$

4.2.  $G(\text{4-pentenal}) = \dfrac{k_{4\cdot15}}{k_{4\cdot15} + k_{4\cdot16}\,[\text{C}_5\text{H}_8]} \cdot G(\text{cyclopentanone}_T)$

$k_{4\cdot16}/k_{4\cdot15} = 21.$

$$\therefore \quad \text{Fraction} = \frac{1}{1 + 21 \times 0\cdot1} = 0\cdot32.$$

4.3.  $\text{C}_2\text{H}_6{}^* \rightarrow \text{CH}_4 + \text{CH}_2$ :

$\Delta H$ for reaction of ground state ethane is given by

$$\Delta H = -74\cdot9 + 398 - (-84\cdot5)$$
$$= 407\cdot6 \text{ kJ mol}^{-1}$$

Ethane molecule must have an excitation energy at least equal to this.

$96\cdot49 \text{ kJ mol}^{-1} = 1 \text{ ev per molecule}$

$$\therefore \quad \text{minimum energy} = \frac{407\cdot6}{96\cdot49} = 4\cdot2 \text{ eV.}$$

5.1.  $$2 \times 10^{-8} = \left(\frac{4D}{6\cdot02 \times 10^{22} \times 3 \times 10^{8}}\right)^{1/2}$$

$$D = 1\cdot81 \times 10^{15} \text{ eV cm}^{-3} \text{ s}^{-1}.$$

5.2.  $pK_{OH} = 11\cdot8 \quad K_{OH} = 1\cdot58 \times 10^{-12}$

Degree of ionization $= \dfrac{[\text{O}^-]}{[\text{O}^-] + [\text{OH}]} = \dfrac{1}{1 + \dfrac{[\text{OH}]}{[\text{O}^-]}}$

but $\dfrac{[\text{OH}]}{[\text{O}^-]} = \dfrac{[\text{H}^+]}{K_{OH}}$

At pH 10, degree $= \dfrac{1}{1 + 10^{-10}/(1\cdot58 \times 10^{-12})} = 0\cdot016.$

At pH 13, degree $= \dfrac{1}{1 + 10^{-13}/(1\cdot58 \times 10^{-12})} = 0\cdot94.$

5.3.  $k(\text{H} + \text{O}_2) = 2\cdot6 \times 10^{10} \text{ M}^{-1} \text{ s}^{-1}$

$k(\text{H} + \text{C}_2\text{H}_5\text{OH}) = 1\cdot5 \times 10^{7} \text{ M}^{-1} \text{ s}^{-1}$

Fraction $= \dfrac{1\cdot5 \times 10^{7}\,[\text{C}_2\text{H}_5\text{OH}]}{1\cdot5 \times 10^{7}\,[\text{C}_2\text{H}_5\text{OH}] + 2\cdot6 \times 10^{10} \times 10^{-3}}$

At $[\text{C}_2\text{H}_5\text{OH}] = 10^{-2}$ M, fraction $= 0\cdot0057$

At $[\text{C}_2\text{H}_5\text{OH}] = 1$ M, fraction $= 0\cdot37.$

5.4.  Dose rate will be approximately $0\cdot642 \times 10^{18} \text{ eV l}^{-1} \text{ s}^{-1}$

$$G(-\text{I}_2) = \frac{7\cdot6}{2} = 3\cdot8$$

Time to consume $10^{-4}$ M $\text{I}_2$

$$= 10^{-4} \times 6\cdot02 \times 10^{23} \times \frac{100}{3\cdot8} \times \frac{1}{0\cdot642 \times 10^{18}} \text{ s}$$

$$= 41 \text{ minutes.}$$

6.1.  $G(-\text{Ce(IV)}) = G_{\text{H}} + G_{\text{e}} + 2G_{\text{H}_2\text{O}_2} - G_{\text{OH}}$

6.2.  1 rad $= 6.24 \times 10^{13}$ eV g$^{-1}$

Total weight of gas $= 100 \times \dfrac{273}{298} \times \dfrac{1}{22.4} \left( \dfrac{28+32}{2} \right)$

$= 122.8$ g

$\therefore$  Consumption of N$_2$

$= 122.8 \times 10^5 \times 6.24 \times 10^{13} \times \dfrac{5}{100} \times \dfrac{28}{6.02 \times 10^{23}} \times 24$

$= 0.043$ g day$^{-1}$

Weight is too small for industrial process.

6.4.  $[C_6H_{14}] = \dfrac{659}{86} = 7.66$ M

$G(H_2)^0 = G_1 + G_2 = 3.16 + 2.12 = 5.28$

$\therefore \quad \dfrac{1}{5.28 - G(H_2)} = \dfrac{1}{3.16} + \dfrac{1}{3.16} \cdot \dfrac{7.66}{5 \times 10^{-3}} \cdot \dfrac{4.9 \times 10^6}{1.05 \times 10^{10}}$

$G(H_2) = 3.44.$

$G(-DPPH) = 2G_1 = 6.32.$

# Bibliography and notes

*General texts*

O'DONNELL, J. H., and SANGSTER, D. F. (1970) *Principles of radiation chemistry.* Arnold, London. A good, concise and up-to-date introduction.

SPINKS, J. W. T. and WOODS, R. J. (1964) *An introduction to radiation chemistry.* Wiley, New York and London. A good and detailed text but somewhat out of date.

SWALLOW, A. J. (1960) *Radiation chemistry of organic compounds.* Pergamon Press, Oxford. A limited but useful introduction.

*Chapter 1.*

CHARLESBY, A. (editor) (1964) *Radiation sources.* Pergamon Press, Oxford. A detailed text.

DORFMAN, L. M. and MATHESON, M. S. (1965) *Pulse radiolysis*, in, *Progress in reaction kinetics*, vol. 3, p. 237. (PORTER, G. editor) Pergamon Press, Oxford. An excellent review article.

HAISSINSKY, M. (1964) *Nuclear chemistry and its applications.* (Trans. TUCK, D. G.) Addison Wesley, Reading, Mass. Essentially a volume on radiochemistry.

KEENE, J. P. (1965) *The technique of pulse radiolysis by optical absorption*, in, *Pulse radiolysis*, p. 1. (BAXENDALE, J. H., EBERT, M., KEENE, J. P., and SWALLOW, A. J. editors) Academic Press, London and New York. A review article for those interested in experimental details.

*Chapter 2.*

ALLEN, A. O. (1961) *The radiation chemistry of water and aqueous solutions*, p. 13. Van Nostrand, New York. A good general survey but a little out of date.

BETHE, H. A. and ASHKIN, J. (1953) in, *Experimental nuclear physics*, vol. 1, p. 166. (Segrè, E. editor) Wiley, New York.

BURNS, W. G. and BARKER, R. (1965) *Dose rate and linear energy transfer effects in radiation chemistry*, in *Progress in reaction kinetics*, vol. 3, p. 303. (Porter, G. editor) Pergamon Press, Oxford. A useful review article.

DAINTON, F. S. (1965) Important early stages in the radiolysis of solid and liquid systems. *Pure appl. Chem.*, **10**, 395. A clearly written review and one which should be read.

WHYTE, G. N. (1959) *Principles of radiation dosimetry.* Wiley, New York. A detailed survey.

*Report of the International Commission on radiological units and measurements*, (1959) (ICRU). Handbook 78 (1961) National Bureau of Statistics (U.S.)

*Chapter 3.*

ANBAR, M. and HART, E. J. (1970) *The hydrated electron.* Wiley, New York. An excellent survey with a wealth of data and one which should be read.

DAINTON, F. S. (1967) *The chemistry of the electron*, in, *Proceedings of the fifth Nobel symposium – fast reactions and primary processes in chemical kinetics*, p. 185. (Claesson, S. editor) Interscience, New York. A useful review.

FREEMAN, G. R. (1970) *Radiation chemistry of ketones and aldehydes*, in, *The chemistry of the carbonyl group*, p. 343. (Zabicky, J. editor) Interscience, London. A good survey of primary processes.

JACKSON, R. A. (1972) *Mechanism; an introduction to the study of organic reactions.* Clarendon Press, Oxford.

LIND, S. C., HOCHANADEL, C. J., and GHORMLEY, J. (1961) *Radiation chemistry of gases.* Reinhold, New York. The standard work on this topic.

*Chapter 4.*

BAXENDALE, J. H. and FITTI, M. (1972) Yield of triplet state benzene in the pulse radiolysis of solutions of some aromatics. *J. chem. Soc., Faraday Transactions II*, **68**, 218. A good source of references.

COMPTON, R. N. and HUEBNER, R. H. (1970) *Collisions of low energy electrons with molecules*, in, *Advances in radiation chemistry.* p. 281. (Burton, M. and Magee, J. L. editors) Wiley, New York. Review article.

JOHNSON, G. R. A. and SCHOLES, G. (editors) (1971) *The chemistry of ionization and excitation.* Taylor and Francis, London. A good survey of the subject.

PHILLIPS, G. O. (editor)(1966) *Energy transfer in radiation processes.* Elsevier, Amsterdam. Useful text but somewhat out of date.

THOMAS, J. K. (1970) Excited states and reactions in liquids. *A. Rev. Phys. Chem.* **21**, 17. A good reference source.

*Chapter 5.*

FESSENDEN, R. W. and SCHULER, R. H. (1970) *Electron spin resonance spectra of radiation produced radicals*, in, *Advances in radiation chemistry*, vol. 2, p. 1. (BURTON, M. and MAGEE, J. L. editors) Wiley, New York. This excellent review contains a wealth of e.s.r. data.

HUGHES, G. (1969) *Effect of high energy radiation*, in, *Comprehensive chemical kinetics*, vol. 3, p. 67. (BAMFORD, C. H. and TIPPER, C. F. H. editors) Elsevier, Amsterdam. A summary of the chief processes occurring.

McLAUCHLAN, K. A. (1972) *Magnetic resonance.* Clarendon Press, Oxford. A concise introduction to e.s.r. (and n.m.r.).

SHERMAN, W. V. (1969) *Free radical intermediates in the radiation chemistry of organic compounds*, in, *Advances in free radical chemistry*, vol. 3, p. 1. (WILLIAMS, G. H. editor) Academic Press, New York. A good general treatment.

*Chapter 6.*

ALLEN, A. O. (1961) *The radiation chemistry of water and aqueous solutions.* Van Nostrand, New York.

BASSON, R. A. (1971) *The radiation chemistry of the hydroxyl group*, in, *The chemistry of the hydroxyl group*, p. 937. (PATAI, S. editor) Interscience, London. A general review.

DRAGANIC, I. G. and DRAGANIC, Z. D. (1971) *Radiation chemistry of water.* Academic Press, New York. A good, up-to-date account which ought to be read.

HUGHES, G. (1965) *Oxidation reactions induced by ionizing radiation*, in, *Oxidation and combustion reviews*, p. 48 (TIPPER, C. F. H. editor) Elsevier, Amsterdam. Useful survey.

SCHULTZ, A. R. (1964) *Radiation and crosslinking by radiation*, in, *Chemical reactions of polymers*, p. 723. (FETTES, E. M. editor) Interscience, New York. General review.

*Chapter 7.*

CHARLESBY, A. (1960) *Atomic radiation and polymers.* Pergamon Press, New York. A competent introduction but somewhat out of date.

TECHNICAL REPORT SERIES (1968) No. 84, *Radiation chemistry and its applications.* International Atomic Energy Agency, Vienna.

WAGNER, C. D. (1969) *Chemical synthesis by ionizing radiation,* in, *Advances in radiation chemistry,* vol. 1, p. 199. (BURTON, M. and MAGEE, J. L. editors) Wiley, New York. A good review of possible industrial uses and should be read.

# Index